John Fair Stoddard, Making of America Project

The American Intellectual Arithmetic

Containing an Extensive Collection of Practical Questions on the General Principles

of Arithmetic

John Fair Stoddard, Making of America Project

The American Intellectual Arithmetic
Containing an Extensive Collection of Practical Questions on the General Principles of Arithmetic

ISBN/EAN: 9783337217778

Printed in Europe, USA, Canada, Australia, Japan

Cover: Foto ©berggeist007 / pixelio.de

More available books at **www.hansebooks.com**

NEW REVISED EDITION.

THE

AMERICAN

INTELLECTUAL ARITHMETIC:

CONTAINING

AN EXTENSIVE COLLECTION OF PRACTICAL QUESTIONS ON THE GENERAL PRINCIPLES OF ARITHMETIC.

WITH

Concise and Original Methods of Solution,

WHICH SIMPLIFY MANY OF THE MOST IMPORTANT RULES IN WRITTEN ARITHMETIC.

BY

JOHN F. STODDARD, A.M.,

AUTHOR OF THE "NORMAL MATHEMATICAL SERIES," ETC.

NEW YORK:

SHELDON & COMPANY,

No. 8 MURRAY STREET.

STODDARD'S SERIES OF ARITHMETICS.

STODDARD'S JUVENILE MENTAL ARITHMETIC..........
    "   INTELLECTUAL   "   ....●
    "   RUDIMENTS OF   "   ....
    "   NEW PRACTICAL   "   ....

SHORT COURSE.

STODDARD'S PRIMARY PICTORIAL ARITHMETIC.........
    "   COMBINATION SCHOOL   "   ....
    "   COMPLETE   "   ....

———◆———

OLNEY'S HIGHER MATHEMATICS.

OLNEY'S INTRODUCTION TO ALGEBRA.................
    "   COMPLETE SCHOOL   "   ....
    "   UNIVERSITY   "   ....
    "   TEST EXAMPLES IN   "   ....
    "   GEOMETRY...........................
    "   TRIGONOMETRY.......................
    "   GEOMETRY AND TRIGONOMETRY, *School Edition.*
    "   "   "   "   *University Ed*
    "   "   "   CALCULUS...................

PREFACE.

HAVING felt the necessity of a more extended and systematic Intellectual Arithmetic for younger, as well as more advanced pupils, I prepared and used in manuscript, in my own school, for a number of years, such a series of questions as I deemed best adapted to the purpose. After observing the superior mental training derived from their use, and the ease with which pupils thus trained comprehended the more advanced branches of mathematics, I venture to submit them to the public in the following pages, hoping that they may prove as useful to other schools as they have to my own.

The rule which I have observed in preparing this work is: *Tell but one thing at a time, and that in its proper place.*

Although in many particulars the work differs from other "Mental" Arithmetics, as an examination of the "questions" will show, mention of these differences is omitted, and the following exposition of its arrangements of subjects is presented.

Chapters First, Second, Third, and *Fourth,* from Lesson I to Lesson XV, treat respectively of Addition, Subtraction, Multiplication, and Division of simple numbers; each of which is rendered familiar by an extensive collection of practical questions. Lesson VII consists of questions which combine Addition and Subtraction; Lesson IX, of questions combining Addition, Subtraction, and Multiplication; Lesson XIII, of questions combining the twelve previous Lessons; and Lesson XIV, of questions in Proportion. Thus, an intimate connection between Lessons and even Chapters is kept up through the entire work, with the exception of *Chapter Fifth,* Lesson XIV to Lesson XXVII, which contains some of the most important Tables of Weights and Measures; each of which is illustrated with appropriate questions.

Chapter Sixth, from Lesson XXVI to Lesson XLVI, is devoted to the subject of Fractions, in which twenty lessons are many original combinations of numbers and concise analyses.

Chapter Seventh, from Lesson XLVI to Lesson LIX, consists of practical and intricate questions of various kinds, which require for their solution a thorough knowledge of the preceding Chapters. This Chapter (perhaps not contained in any similar work), when understood, will be of great benefit to those who are studying, or who intend to study Algebra.

Chapter Eighth, from Lesson LIX to the end, includes Interest, Discount, and Per Cent., in their various modifications. The method of treating these subjects is original ; and renders the rules under these heads, in Written Arithmetics (which are often incomprehensible to pupils), perfectly intelligible, by reducing the whole to one continued train of reasoning.

This Chapter, thoroughly taught, can not fail to quicken, strengthen, and develop the reasoning powers. Bringing into exercise, as thorough teaching of it will, nearly every principle taught in the twenty lessons of *Chapter Sixth,* and also the greater part of *Chapter Seventh,* the pupil will acquire the habit of systematically classifying his knowledge, and be enabled to call to his aid such portions of it as will assist in illustrating or demonstrating the subject under consideration.

That Intellectual Arithmetic, when properly taught, is better calculated, than any other study, to invigorate and develop the reasoning faculties of the mind, to produce accurate and close discrimination, and to enable the pupil to acquire a knowledge of the Higher Mathematics with greater ease, can scarcely admit of a doubt.

J. F. STODDARD.

New York, August 1, 1860.

Publishers' Note.—A new edition of this popular Intellectual Arithmetic, carefully revised by the author, is here presented in new and larger type, and on larger pages, without any changes which might interfere with its use in the same classes with previous editions. The "Lessons" are numbered, in regular order, throughout the book. In Lesson XXVI are full Tables of Metrical Weights and Measures on the Decimal System, the simplicity of which is an important consideration for instructors of youth.

Prof. Stoddard's new Key to this book, containing his Methods of Teaching Intellectual Arithmetic, is now published

SUGGESTIONS TO TEACHERS.

For the benefit of those whose experience in teaching Mental Arithmetic is limited, the following suggestions are made of such methods of teaching this important subject, as may prove best suited to fix the attention, strengthen the memory, develop the reasoning powers, and secure rapid and accurate computation.

One thing at a time should be taught, thoroughly, and in its proper order.

Recitations and exercises for children should be short, and during their continuance the careful attention of each member of the class should be secured, and thereby animation and promptness will be encouraged.

The lesson should be assigned previous to recitation, to afford the pupils an opportunity for an examination and study of it; and during class exercise, pupils should not use the book.

Drills, Illustrations, and *Explanations* should occupy at least *one half* of the time devoted to each recitation for children.

Care should be taken that the positions of children should be good, and that the language used be strictly correct in articulation, pronunciation, and construction, and addressed to the person asking the question. Both listlessness and hurried solutions should be avoided; in the latter, pupils not unfrequently pronounce *and, if, what, costs, quarts,* as follows, *an, ef, wat, coss, quats.* By careful attention to these particulars, lessons in Intellectual Arithmetic will be valuable exercises in address, elocution, grammar, rhetoric, and logic, and pupils will acquire both a ready command of their thoughts, and a fluency of language in expressing them.

A Question should be read slowly and distinctly, and a pupil be required to repeat it accurately, and analyze it thoroughly, according to the forms given. There should be no interruption, except when the teacher deems it necessary to make a correction or an important criticism.

Pupils should be called upon promiscuously, and not in rotation, to take part in the recitation.

Class Drills should not be employed as a regular method of recitation, but simply to fix in the minds of pupils such tabular facts as can more readily be learned by concert recitation, to enliven the exercises, to give animation to the class, and confidence to the timid pupils.

Combinations of figures in the Tables should be thoroughly learned by the pupil, and both rapidity and accuracy should characterize his operations in Addition before he is required to study other parts of the subject. As aids to this result, exercises such as the following may be made: the teacher may give out the number 6, and require each member of the class to write on his slate the various combinations of two numbers, with the proper signs, that will produce the given number. Most pupils will at first perform this exercise in an irregular manner, and the teacher should instruct them, by forming the proper arrangement on the blackboard and explaining it.

PUPIL'S ARRANGEMENT.	TEACHER'S ARRANGEMENT.
$2+4=6$	$5+1=6$ $\qquad$ $6=1+5$
$1+5=6$	$4+2=6$ $\qquad$ $6=2+4$
$4+2=6$	$3+3=6$ or, $6=3+3$
$5+1=6$	$2+4=6$ $\qquad$ $6=4+2$
$3+3=6$	$1+5=6$ $\qquad$ $6=5+1$

Similar exercises may be had on Subtraction, Multiplication, and Division, with the appropriate signs.

Every combination of this arrangement may be illustrated with objects, such as pebbles, grains of corn, beans, so that the class may clearly understand each.

A pupil thoroughly drilled in the fundamental operations of Arithmetic, will not only be able to perform them with facility and accuracy, but will have made great progress toward an easy and complete mastery of the Science of Numbers.

The advantages of this exercise are, that it insures a thorough knowledge of the Tables; it teaches to write figures and signs properly; it gives pleasant employment to the bodies and minds of pupils, and therefore helps to secure good order.

More explanations on Methods of Teaching Intellectual Arithmetic, are presented in the new Key to this work, which book may be a valuable assistant to those who have found difficulties in using Stoddard's Intellectual Arithmetic.

ARITHMETIC.

ADDITION.

Addition is the process of uniting like numbers into one sum.

The Sum, or Amount, is a number equal to all the numbers added.

LESSON I.

1. 2 and 1 are how many?

ANALYSIS.—Two and one are three.

2. 2 and 2 are how many?
3. 2 and 3 are how many?
4. 2 and 4 are how many?
5. 2 and 5 are how many?
6. 2 and 6 are how many?
7. 2 and 7 are how many?
8. 2 and 8 are how many?
9. 2 and 9 are how many?
10. 3 and 2 are how many?
11. 3 and 3 are how many?
12. 3 and 4 are how many?
13. 3 and 5 are how many?
14. 3 and 6 are how many?
15. 3 and 7 are how many?
16. 3 and 8 are how many?
17. 3 and 9 are how many?
18. 4 and 3 are how many?
19. 4 and 4 are how many?
20. 4 and 5 are how many?
21. 4 and 6 are how many?
22. 4 and 7 are how many?
23. 4 and 8 are how many?
24. 4 and 9 are how many?

25. James killed 2 birds, and John killed 1 bird; how many birds did both kill?

> ANALYSIS.—If James killed 2 birds and John 1, they together killed 2 birds and 1 bird, which are 3 birds.

26. I gave 2 cents to Henry, and 2 cents to Harvey; how many cents did both receive?

27. Hiram had 2 cents, and his brother gave him 3 cents more; how many cents had he then?

28. George gave me 2 apples, and Mary gave me 4 apples; how many apples did both give me?

29. A man had 2 cows, and he purchased 5 cows more; how many cows had he then?

30. John's father gave him 2 oranges, and his mother gave him 6; how many did he receive?

31. Philo bought 2 peaches, and his brother gave him 7; how many peaches had he then?

32. Philip gave me 2 plums, and Myron gave me 8; how many plums did they together give me?

33. A farmer had 2 horses, and bought 9 more; how many horses had he then?

34. William had 3 oranges, and Moses gave him 2 more; how many had he then?

35. John bought 3 apples, and I gave him 3; how many had he then?

36. Philip paid 3 cents for some nuts, and 4 cents for some candy; how many cents did he pay for both?

37. I paid 5 cents for some paper, and 3 cents for a stamp; how much did I pay for both?

38. A merchant bought 3 barrels of sugar and 6 barrels of molasses; how many barrels did he buy?

39. Ralph is 3 years old, and Edward is 7; what is the sum of their ages?

40. A lemon cost 3 cents, and a pine-apple cost 8; what sum did both cost?

41. James solved 3 questions in arithmetic, and Oliver 9; how many did both solve?

42. If it take 4 yards of cloth for a coat, and 1 yard of cloth for a vest, how many yards will it take for both ?

43. Samuel bought 4 marbles, and found 4 ; how many marbles had he then ?

44. Isaac bought 4 sheets of paper, and I gave him 5 ; how many had he then ?

45. A man bought an apple for 4 cents, and a pear for 6 cents ; how much did the apple and pear together cost ?

46. If Mary has 4 books, and her father should give her 7, how many books would she then have ?

47. William has 4 marbles in his hand, and 8 in his pocket ; how many marbles has he in all ?

48. Charles walked 4 miles, and rode 9 ; how many miles did he go ?

49. In a certain class there are 5 boys, and 4 girls ; how many pupils are in the class ?

LESSON II.

1.	5	and	4	are	how	many ?
2.	5	and	5	are	how	many ?
3.	5	and	6	are	how	many ?
4.	5	and	7	are	how	many ?
5.	5	and	8	are	how	many ?
6.	5	and	9	are	how	many ?
7.	6	and	5	are	how	many ?
8.	6	and	6	are	how	many ?
9.	6	and	7	are	how	many ?
10.	6	and	8	are	how	many ?
11.	6	and	9	are	how	many ?
12.	7	and	6	are	how	many ?
13.	7	and	7	are	how	many ?
14.	7	and	8	are	how	many ?
15.	7	and	9	are	how	many ?

16.	.8	and	7	are	how	many ?
17.	8	and	8	are	how	many ?
18.	8	and	9	are	how	many ?
19.	8	and	5	are	how	many ?
20.	9	and	6	are	how	many ?
21.	9	and	8	are	how	many ?
22.	9	and	9	are	how	many ?
23.	9	and	10	are	how	many ?
24.	9	and	7	are	how	many ?
25.	9	and	11	are	how	many ?

26. Mary answered 5 questions correctly, and 4 incorrectly; how many questions did she answer?

27. A beggar met two boys; one gave him 5 cents, and the other gave him 6 cents; how many cents did both give him?

28. A man bought a hat for 5 dollars, and a pair of boots for 6 dollars; how much did he pay?

29. There are 9 boys on one bench, and 8 on another; how many are on both?

30. Maria gave her teacher 5 pinks and 7 roses; how many flowers did she give him?

31. Harry caught 5 squirrels, and Henry caught 8; how many were caught by both?

32. If we learn 5 pages this week, and 9 next, how many shall we learn in the two weeks?

33. Frank sold a melon for 6 cents, and an orange for 5 cents; for how many cents did he sell both?

34. John bought 6 whips, and Joseph gave him 6; how many had John then?

35. George had 6 chestnuts, and Richard gave him 7; how many had George then?

36. Henry bought 6 figs, and Sarah bought 8; how many were bought by both?

37. Rebecca has 6 oranges, and Catherine has 9; how many oranges have both?

38. A boy bought 7 apples, and his father gave him 6; how many had he then?

39. Minerva bought 7 yards of ribbon, and her mother gave her 7; how many yards had Minerva then?
40. There were 7 boys sitting on one bench, and 8 on another; how many boys were on both?
41. There were 7 boys at play, and 9 other boys joined them; how many boys were in all?
42. If I have 8 cents in one hand, and 7 cents in the other, how many have I in both hands?
43. If Mary has 8 peaches, and Margaret has 9, how many have both?
44. Sally gave 9 cents for some thread, and 7 cents for some needles; how much did the needles and thread cost?
45. Charles has 9 marbles, and Albert has 5; how many marbles have Charles and Albert together?
46. 9 birds were in a tree, and 6 were on the ground; how many birds in all?
47. Sarah gave 9 cents for cinnamon, and 7 cents for raisins; how many cents did both cost?
48. George shot 9 pigeons, and James shot 8; how many did both shoot?
49. Russel caught 7 fish, and Robert caught 5; how many did both catch?
50. In one field there are 8 horses, and in another there are 9; how many are there in both?

———⊶⊙⊶———

LESSON III.

1. How many are 10 and 2?   10 and 3?   10 and 4?
   10 and 5?   10 and 6?   10 and 7?   10 and 9?
   10 and 8?   10 and 10?
2. How many are 2 and 2?   2 and 12?   2 and 22?
   2 and 32?   2 and 42?   2 and 52?   2 and 62?
   2 and 72?   2 and 82?   2 and 92?

3. How many are 3 and 3?   3 and 13?   3 and 23?
  3 and 33?   3 and 43?   3 and 53?   3 and 63?
  3 and 73?   3 and 83?   3 and 93?   96 and 4?

4. How many are 4 and 4?   4 and 14?   4 and 24?
  4 and 34?   4 and 44?   4 and 54?   4 and 64?
  4 and 74?   4 and 84?   4 and 94?   98 and 2?

5. How many are 5 and 5?   5 and 15?   5 and 25?
  5 and 35?   5 and 45?   5 and 55?   5 and 65?
  5 and 75?   5 and 85?   5 and 95?

6. How many are 6 and 6?   6 and 16?   6 and 26?
  6 and 36?   6 and 46?   6 and 56?   6 and 66?
  6 and 76?   6 and 86?   6 and 96?

7. How many are 7 and 7?   7 and 17?   7 and 27?
  7 and 37?   7 and 47?   7 and 57?   7 and 67?
  7 and 77?   7 and 87?   7 and 97?

8. How many are 8 and 8?   8 and 18?   8 and 28?
  8 and 38?   8 and 48?   8 and 58?   8 and 68?
  8 and 78?   8 and 88?   8 and 98?

9. How many are 9 and 9?   9 and 19?   9 and 29?
  9 and 39?   9 and 49?   9 and 59?   9 and 69?
  9 and 79?   9 and 89?   9 and 99?

10. How many are 10 and 11?   10 and 21?   10 and
31?   10 and 41?   10 and 51?   10 and 61?   10
and 71?   10 and 81?   10 and 91?

11. How many are 10 and 12?   10 and 22?   10 and
32?   10 and 42?   10 and 52?   10 and 62?   10
and 72?   10 and 82?   10 and 92?

12. How many are 10 and 4?   10 and 14?   10 and
24?   10 and 34?   10 and 44?   10 and 54?   10
and 64?   10 and 74?   10 and 84?   10 and 94?

13. How many are 11 and 3?   11 and 13?   11 and
23?   11 and 33?   11 and 43?   11 and 53?   11
and 63?   11 and 73?   11 and 83?   11 and 93?

14. How many are 11 and 4?   11 and 14?   11 and
24?   11 and 34?   11 and 44?   11 and 54?   11
and 64?   11 and 74?   11 and 84?   11 and 94?

15. How many are 10 and 5?   10 and 15?   10 and

35? 10 and 45? 10 and 55? 10 and 65? 10 and 75? 10 and 85? 10 and 95? 10 and 25?

16. How many are 11 and 5? 11 and 15? 11 and 25? 11 and 35? 11 and 45? 11 and 55? 11 and 65? 11 and 75? 11 and 85? 11 and 95?

17. What is the sum of 3 and 8? 3 and 18? 3 and 28? 3 and 38? 3 and 48? 3 and 58? 3 and 68? 3 and 78? 3 and 88? 3 and 98?

18. What is the sum of 8 and 4? 8 and 14? 8 and 24? 8 and 34? 8 and 44? 8 and 54? 8 and 64? 8 and 74? 8 and 84? 8 and 94?

19. What is the sum of 7 and 5? 7 and 17? 7 and 27? 7 and 37? 7 and 47? 7 and 57? 7 and 67? 7 and 77? 7 and 87? 7 and 97?

20. What is the sum of 8 and 6? 8 and 16? 8 and 26? 8 and 36? 8 and 46? 8 and 56? 8 and 66? 8 and 76? 8 and 86? 8 and 96?

LESSON IV.

1.	8	and	9	are how many?
2.	11	and	7	are how many?
3.	10	and	9	are how many?
4.	7	and	14	are how many?
5.	6	and	12	are how many?
6.	9	and	15	are how many?
7.	11	and	18	are how many?
8.	15	and	12	are how many?
9.	14	and	13	are how many?
10.	16	and	14	are how many?
11.	21	and	12	are how many?
12.	24	and	13	are how many?
13.	25	and	15	are how many?
14.	27	and	13	are how many?
15.	23	and	17	are how many?

16.	29	and	11	are	how	many?
17.	30	and	20	are	how	many?
18.	34	and	15	are	how	many?
19.	32	and	18	are	how	many?

The Sign of Addition is +, and is called *plus*.

When + is placed between two numbers, it denotes that they are to be added together.

The Sign of Equality is =, and is read *equals*, or *equal to*.

When = is placed between two numbers, it denotes that they are equal to each other.

Thus, 2+3=5, is read *two plus three equals five.*

* * + * * * = * * * * *

20.	$33 + 44$ are how many?
21.	$35 + 15$ are how many?
22.	$36 + 12$ are how many?
23.	$40 + 36$ are how many?
24.	$40 + 29$ are how many?
25.	$44 + 20$ are how many?
26.	$48 + 32$ are how many?
27.	$45 + 35$ are how many?
28.	$4 + 8 + 6 =$ what number?
29.	$8 + 2 + 7 =$ what number?
30.	$10 + 7 + 3 =$ what number?
31.	$12 + 10 + 9 =$ what number?
32.	$15 + 12 + 6 =$ what number?
33.	$18 + 4 + 10 =$ what number?
34.	$24 + 16 + 12 =$ what number
35.	$22 + 33 + 11 =$ what number?
36.	$15 + 16 + 2 =$ what number?
37.	$28 + 12 + 15 =$ what number?
38.	$46 + 24 + 19 =$ what number?
39.	$12 + 8 + 6 + 4 =$ what number?
40.	$24 + 10 + 6 + 12 =$ what number?
41.	$22 + 32 + 6 + 10 =$ what number?
42.	$37 + 23 + 15 =$ what number?
43.	$62 + 26 + 12 + 8 =$ what number?

LESSON V.

1. Three boys, James, Joseph, and Jacob, gave some money to a beggar; James gave him 6, Joseph 8, and Jacob 10 cents; how many cents did all give him?
2. Gave 8 cents to John, 4 cents to Morgan, and 2 cents to Samuel; how many cents did all receive?
3. Henry has 3 marbles, Harvey has 10, and Charles has 7; how many marbles have all?
4. Gave 7 nuts to one boy, 6 to another, and 7 to another; how many nuts did I give to the three boys?
5. Bought a basket of strawberries for 7 cents, a basket of cherries for 4 cents, and a basket of plums for 8 cents; how many cents did all cost?
6. Lydia has 9 pinks, Mary 10, and Ann 7; how many pinks have all?
7. Bought a knife for 14 cents, and a ball for 12 cents; how much did both cost?
8. Gave 18 cents for an arithmetic, 2 for a pencil, and 10 for a slate; how much did all cost?
9. James had 12 cents, and his mother gave him 13 more; how many had he then?
10. Robert shot 9 birds, Richard shot 11, and James shot 12; how many did all shoot?
11. A boy bought a pound of butter for 44 cents, a pound of meat for 20 cents, and a bunch of lettuce for 7 cents; how many cents did all cost?
12. Bought a pound of raisins for 10 cents, a pound of candy for 12 cents, and a pound of cinnamon for 15 cents; how much was the whole cost?
13. John had 20 marbles, Matthew 9, and Morgan 12; how many had they in all?
14. James bought a pigeon for 9 cents, a robin for 10 cents, and a squirrel for 12 cents; how much did all cost?
15. A lady bought some pins for 15 cents, some

thread for 10 cents, and some lace for 18 cents; how many cents did all these articles cost?

16. A gentleman bought a hat for 6 dollars, a vest for 5 dollars, and a coat for 20 dollars; how many dollars did he pay for all?

17. A man bought a watch for 40 dollars, a chain for 15 dollars, and a gold pen for 5 dollars; how much did he pay for these three articles?

18. Jackson gave 25 cents to his sister, and 23 to his mother; how many cents did he give away?

19. Bought a barrel of flour for 7 dollars, a barrel of pork for 12 dollars, and a barrel of fish for 11 dollars; how much was the whole cost?

20. Bought a horse for 60 dollars, a cow for 20 dollars, and a colt for 25 dollars; how much did all cost?

21. If your father should give you 12 cents, your mother 14 cents, and your sister 4 cents, how many cents would you then have?

22. A boy spent 11 cents for candy, 9 cents for a ball, and 5 cents for a top; how many cents did he spend for all?

23. A man bought some butter for 57 dollars, and some molasses for 23 dollars; how many dollars did he pay for both?

24. A boy traveled 17 miles one day, and 23 the next; how far did he travel in the two days?

25. A lady bought a hat for 7 dollars, a dress for 9 dollars, and a gold watch for 60 dollars; how many dollars did all cost?

26. A mechanic sold a wagon for 57 dollars, and a sleigh for 43 dollars; how much did he receive for both?

27. A boy saw 24 pigeons on one tree, and 36 on another; how many did he see in all?

28. In a certain recitation 21 questions were answered correctly, and 9 incorrectly; how many questions were asked during the recitation?

29. Gave 37 dollars for a chaise, 2 dollars for a whip, and 11 dollars for a buffalo robe; how much was given for all?

30. If a horse is worth 60 dollars, and a sleigh 75 dollars, what is the value of both?

31. Matthew is 15 years old, Morgan is 7, and Martin is 10; what is the sum of their ages?

32. A man bought a load of hay for 7 dollars, a load of rye for 36 dollars, and a load of wheat for 57 dollars; how much was the whole cost?

33. A man is 48 years old, and his wife is 32 years old; what is the sum of their ages?

34. A farmer bought a horse for 60 dollars, and a yoke of oxen for 75 dollars; how much did the horse and oxen together cost?

35. John gave 11 apples to his brother, 9 to his sister, and kept 12 himself; how many apples had he at first?

36. Simeon hoed 12 rows of corn, Simon 15, James 13, and John 11; how many rows did they together hoe?

37. A merchant sold 30 barrels of flour one week, 37 the next week, and 33 the following week; how many barrels did he sell during the three weeks?

38. A merchant sold a barrel of sugar for 25 dollars, a barrel of rum for 15 dollars, and a hogshead of molasses for 23 dollars; how much did he receive for all these articles?

39. A man bought a firkin of butter for 9 dollars, a keg of molasses for 7 dollars, a box of cheese for 4 dollars, and a box of raisins for 5 dollars; how much was the entire cost?

40. A lady bought a silk dress for 48 dollars, a muff for 22 dollars, a shawl for 17 dollars, and a pair of gloves for 1 dollar: the entire cost is required.

2*

SUBTRACTION.

Subtraction is the process of finding the difference between two numbers.

The Terms in Subtraction are three, the Minuend, Subtrahend, Remainder.

1. *The Minuend* is the number to be *diminished;*

2. *The Subtrahend* is the number to be *subtracted;*

3. *The Remainder,* or *Difference,* is the excess of the minuend over the subtrahend.

LESSON VI.

1. If I have 3 apples, and give 1 of them to Richard, how many shall I have left ?

> ANALYSIS.—If I have 3 apples, and give 1 to Richard, I shall have remaining the difference between 3 apples and 1 apple, which is 2 apples.

2. William had 4 chestnuts, and gave 1 to his brother; how many had he left ?

3. Martha had 5 books, and on her way to school, lost one of them; how many had she left ?

4. Cornelia had 6 apples, and gave 1 to her brother; how many had she left ?

5. Rachel had 10 pins, and lost 1 of them; how many had she left ?

6. Martha had 12 pears, and gave 2 to Elizabeth; how many had she left ?

7. Jane had 5 canary-birds, and gave 2 of them to Eliza; how many had Jane left ?

8. James had 6 apples, and gave 2 away; how many had he remaining ?

9. Cornelia learned 12 letters yesterday, and has forgotten 3 of them; how many does she remember ?

10. James had 10 marbles, and lost 3; how many had he left ?

11. Mary found 9 roses on her bush, and picked off 4 of them; how many remained on the bush?

12.	How	many	are	4	less	2?
13.	How	many	are	5	less	3?
14.	How	many	are	7	less	3?
15.	How	many	are	9	less	4?
16.	How	many	are	9	less	3?
17.	How	many	are	9	less	7?
18.	How	many	are	9	less	5?
19.	How	many	are	11	less	4?
20.	How	many	are	10	less	4?
21.	How	many	are	11	less	5?
22.	How	many	are	14	less	4?
23.	How	many	are	8	less	5?
24.	How	many	are	13	less	3?
25.	How	many	are	14	less	5?
26.	How	many	are	17	less	7?
27.	How	many	are	15	less	5?
28.	How	many	are	13	less	10?
29.	How	many	are	23	less	3?
30.	How	many	are	27	less	7?

The Sign of Subtraction is −, and is called *minus.*

When − is placed between two numbers, it indicates that the number on the right is to be subtracted from the number on the left. Thus, 7−5=2, indicates that 5 is to be subtracted from 7, and that the Remainder, or Difference, is 2; and is read 7 *minus* 5 *equals* 2.

31.	$8 - 5 =$ how many?
32.	$9 - 7 =$ how many?
33.	$10 - 8 =$ how many?
34.	$11 - 8 =$ how many?
35.	$12 - 6 =$ how many?
36.	$13 - 8 =$ how many?
37.	$14 - 8 =$ how many?
38.	$18 - 8 =$ how many?
39.	$22 - 12 =$ how many?
40.	$24 - 14 =$ how many?

41.	12 − 4 = how many ?
42.	28 − 8 = how many ?
43.	20 − 5 = how many ?
44.	20 − 8 = how many ?
45.	20 − 9 = how many ?
46.	20 − 7 = how many ?
47.	20 − 10 = how many ?
48.	20 − 15 = how many ?
49.	24 − 10 = how many ?
50.	25 − 10 + 5 = what number ?
51.	26 − 10 + 4 = what number ?
52.	28 − 10 + 2 = what number ?
53.	27 − 10 + 5 = what number ?
54.	29 − 10 + 6 = what number ?
55.	32 − 10 + 8 = what number ?
56.	34 − 10 + 7 = what number ?
57.	36 − 10 + 8 = what number ?
58.	35 − 10 + 4 = what number ?
59.	37 − 10 + 7 = what number ?
60.	38 − 10 + 8 = what number ?

61. 39 − 10 + 9 = what number ?
62. 47 − 10 + 6 = what number ?
63. 40 − 12 + 9 = what number ?
64. 42 − 20 + 7 = what number ?
65. 45 − 20 + 5 = what number ?
66. 46 − 20 + 6 = what number ?
67. 47 − 20 + 8 = what number ?
68. 47 − 37 + 4 = what number ?
69. 49 − 19 + 9 = what number ?

70. 52 − 22 + 10 = what number ?
71. 54 − 34 + 11 = what number ?
72. 56 − 46 + 12 = what number ?
73. 57 − 27 + 14 = what number ?
74. 58 − 48 + 9 = what number ?
75. 62 − 30 + 10 = what number ?
76. 65 − 40 + 15 = what number ?
77. 68 − 48 + 16 = what number ?
78. 74 − 34 + 15 = what number ?

79. Gave 7 cents for a spool of thread, and 4 cents for a lemon ; how much more did the thread cost than the lemon ?

80. Paid 18 cents for a pound of butter, and 8 cents for a pound of meat ; how much more was paid for the butter than for the meat ?

81. James bought 18 candles, and gave John 7 of them ; how many had he left ?

82. Sold a quantity of wool for 27 dollars, and received in part payment a barrel of flour worth 5 dollars; how many dollars remain due?

83. James has 27 marbles, and John has 17; how many more has James than John?

84. Harry is 15 years old, and Charles is 9 years old; how many years older is Harry than Charles?

85. A teacher being asked how many pupils he had, answered that he usually had 37, but at present he had only 27; how many were absent?

86. A man purchased a watch for 37 dollars, but found he had only 24 dollars with him; how much must he borrow to pay the balance?

87. Mr. A has 94 sheep, and B has 44; how many more sheep has A than B?

88. Morgan gave 23 cents for some cake, and 14 cents for some cinnamon; how much more did the cake cost than the cinnamon?

89. Michael had 29 cents, and lost 14; how many had he left?

90. In a certain recitation 47 questions were asked, and 9 of them were answered incorrectly; how many were correctly answered?

91. A man sold 23 sheep from a flock consisting of 93; how many sheep remained?

92. Mr. B bought a horse for 35 dollars, and sold it for 46 dollars; how much did he gain?

93. A cow was bought for 25 dollars, and sold for 19 dollars; how much was the loss?

94. A merchant bought a quantity of goods for 95 dollars, but being damaged was obliged to sell them for 80 dollars; how much did he lose?

95. The minuend is 57, the subtrahend is 27; what is the remainder?

96. The minuend is 67, the remainder is 20; what is the subtrahend?

97. The subtrahend is 12, the remainder is 18; what is the minuend?

LESSON VII.

ADDITION AND SUBTRACTION COMBINED.

1. James has 7 chestnuts, and Mary has 4; how many more has James than Mary; and how many have both?

   ANALYSIS.—If James has 7 chestnuts, and Mary has 4, James has as many more than Mary as the difference between 7 chestnuts and 4 chestnuts, which is 3 chestnuts. Together they have 7 chestnuts and 4 chestnuts, which are 11 chestnuts.

2. Bought a barrel of fish for 8 dollars, and some quinces for 3 dollars; how much more did the fish cost than the quinces? What was the cost of both?

3. Gave 15 dollars for a cow and 6 dollars for a sheep; how much more was given for the cow than for the sheep? How much was given for both?

4. Phineas gave 50 cents for a grammar, and 25 cents for an arithmetic; how much was the cost of both? How much did one cost more than the other?

5. Paid 15 dollars for a barrel of rum, and 6 dollars for a barrel of flour; how much was the cost of both; and how much more did the rum cost than the flour?

6. Sold a firkin of butter for 10 dollars, a cheese for 5 dollars, and received in part payment a barrel of flour worth 6 dollars; how much remains due?

7. James gave 12 cents for oranges, 15 cents for cake, and had 13 cents remaining; how much had he at first?

8. Mary bought a comb for 10 cents, a spool of thread for 12 cents, and a paper of needles for 8 cents; she handed the clerk 37 cents; how much change ought she to receive?

9. A man sold a cow for 20 dollars, a calf for 4 dollars, and a sheep for 3 dollars; and in part payment received a wagon worth 17 dollars; how much remains due?

10. A lady bought a ribbon for 24 cents, some tape for 8 cents, and some thread for 12 cents; she had only 60 cents; how much remained after she paid for these articles?

11. Stephen, at one game of marbles, won 4, and at another he lost 6, and then had only 8 remaining; how many had he at first?

12. Sampson having 9 apples, gave 4 to his mother, and 3 to his sister; for his generosity his father gave him 13 more; how many had he then?

13. A man bought some cloth for 12 dollars, and sold it for 18 dollars; how much was his gain?

14. A farmer bought a horse for 63 dollars, and exchanged it for a yoke of oxen; these he sold for 87 dollars; how much did he gain by the operation?

15. A man bought a yoke of oxen for 97 dollars, their services earned for him 40 dollars, and their keeping cost him 13 dollars; he then sold them for 80 dollars; how much did he gain?

16. A box of raisins was bought for 3 dollars, a firkin of butter for 15 dollars, and were both sold for 20 dollars; how much was gained?

17. A farmer sold a cow for 29 dollars, which was 5 dollars more than she cost; how much did she cost?

18. A drover bought some sheep for 40 dollars, some cattle for 130 dollars, and sold them all for 200 dollars; how much was his gain?

19. A jeweller bought a watch for 20 dollars, a chain for 10 dollars, a key for 2 dollars, and sold them all for 42 dollars; how much did he gain by the bargain?

20. $24+12+9=$ how many?

21. $10+30+15=$ how many?

22. $14+16+11=$ how many?

23. $36+9-12=$ how many?

24. $38+22-15=$ how many?

25. $43+37-20=$ how many?

26. $13+26-25=$ how many?

27. $44-20+10-12=$ how many?
28. $27+23-20+ 2=$ how many?
29. $15+25-30+15=$ how many?
30. $20+40-30+10=$ how many?

31. A boy bought a ball for 6 cents; for how much must he sell it to gain 4 cents?

32. A merchant bought a hogshead of molasses for 47 dollars, and paid 3 dollars for cartage; for how much must he sell it to gain 12 dollars?

33. A grocer bought a hogshead of sugar for 30 dollars; for what must he sell it to gain 18 dollars?

34. A drover bought sheep as follows: of one man he bought 24, of another 8, and of another 22: he then sold 20 of them; how many remained unsold?

35. A watch cost 40 dollars; how must it be sold to gain 13 dollars?

36. Four boys bought a melon; one gave 3 cents, another 4, another 8, and the other 6; how much did they pay for the melon?

37. Mary bought 16 plums at one store, and 13 at another; on her way home she ate 11 of them; how many had she left?

38. Matthew had 9 nuts, Mary gave him 10 more, and John gave him enough to make his number 39; how many did John give him?

39. A farmer had 25 sheep in one field, and 15 in another; he then bought as many more as made his number 56; how many did he buy?

40. John has 34 marbles, and Albert 25; how many have both? How many more has John than Albert?

41. A butcher has 57 sheep, and 44 lambs; how many more sheep has he than lambs? How many sheep and lambs together?

42. Paid 97 dollars for a quantity of sugar, and 43 dollars for some molasses; how much more did the sugar cost than the molasses? How much did both cost?

MULTIPLICATION.

Multiplication is the process of taking one number as many times as there are units in another.

Multiplication is also a method of finding the sum of several equal numbers.

The Terms in Multiplication are, the Multiplicand, Multiplier, and Product.

1. *The Multiplicand* is the number to be taken;

2. *The Multiplier* is the number which shows how many times the multiplicand is to be taken;

3. *The Product* is the result obtained.

LESSON VIII.

1.	Two	times	1	are	how	many?
2.	Two	times	2	are	how	many?
3.	Two	times	3	are	how	many?
4.	Two	times	4	are	how	many?
5.	Two	times	5	are	how	many?
6.	Two	times	6	are	how	many?
7.	Two	times	7	are	how	many?
8.	Two	times	8	are	how	many?
9.	Two	times	9	are	how	many?
10.	Two	times	10	are	how	many?
11.	Two	times	11	are	how	many?
12.	Two	times	12	are	how	many?

13. What will 2 oranges cost, at 3 cents each?

ANALYSIS.—If 1 orange cost 3 cents, 2 oranges will cost two times 3 cents; which are 6 cents.

14. What will 2 peaches cost, at 2 cents each?
15. What will 2 apples cost, at 3 cents each?
16. What will 2 pine-apples cost, at 8 cents each?
17. What will 2 pencils cost, at 8 cents each?
18. What will 2 quails cost, at 11 cents each?
19. What will 2 primers cost, at 12 cents each?

2

20. What will 2 citrons cost, at 10 cents each?
21. What will 2 tops cost, at 9 cents each?
22. What will 2 lemons cost, at 4 cents each?

23.	Three	times	2	are	how	many?
24.	Three	times	3	are	how	many?
25.	Three	times	4	are	how	many?
26.	Three	times	5	are	how	many?
27.	Three	times	6	are	how	many?
28.	Three	times	7	are	how	many?
29.	Three	times	-8	are	how	many?
30.	Three	times	9	are	how	many?
31.	Three	times	10	are	how	many?
32.	Three	times	11	are	how	many?
33.	Three	times	12	are	how	many?
34.	Four	times	3	are	how	many?
35.	Four	times	4	are	how	many?
36.	Four	times	5	are	how	many?
37.	Four	times	6	are	how	many?
38.	Four	times	7	are	how	many?
39.	Four	times	8	are	how	many?
40.	Four	times	9	are	how	many?
41.	Four	times	10	are	how	many?
42.	Four	times	11	are	how	many?
43.	Four	times	12	are	how	many?

44. What will 3 quarts of cherries cost, at 6 cents a quart?
45. What will 3 lead pencils cost, at 5 cents each?
46. What will 3 quarts of milk cost, at 4 cents a quart?
47. What will 3 yards of ribbon cost, at 7 cents a yard?
48. What will 4 quarts of chestnuts cost, at 6 cents a quart?
49. What will 4 yards of edging cost, at 5 cents a yard?
50. What will 3 ounces of snuff cost, at 8 cents an ounce?
51. What will 4 ounces of cinnamon cost, at 7 cents an ounce?

52. What will 3 pounds of cheese cost, at 10 cents a pound ?
53. What will 4 sheets of wadding cost, at 8 cents a sheet ?
54. What will 3 yards of calico cost, at 11 cents a yard ?
55. What will 4 skeins of silk cost, at 9 cents a skein ?
56. What will 3 yards of ribbon cost, at 12 cents a yard ?
57. What will 4 pounds of starch cost, at 12 cents a pound ?
58. What will 4 candlesticks cost, at 11 cents each ?
59. What will 4 tops cost, at 10 cents each ?
60. What will 5 apples cost, at 4 cents each ?

The Sign of Multiplication is ×, and is read *multiplied by*, or *times*.

When × is placed between two numbers, it denotes that either is to be multiplied by the other. Thus, $4 \times 5 = 20$, denotes that 4 is to be multiplied by 5 (or 5 by 4), and that the product is 20; and is read 4 *times* 5 *equals* 20, or 4 *multiplied by* 5 *equals* 20.

61. $5 \times 6 =$ how many ?
62. $5 \times 7 =$ how many ?
63. $5 \times 9 =$ how many ?
64. $5 \times 8 =$ how many ?
65. $5 \times 5 =$ how many ?
66. $5 \times 10 =$ how many ?
67. $5 \times 12 =$ how many ?
68. $5 \times 11 =$ how many ?
69. $6 \times 6 =$ how many ?
70. $6 \times 8 =$ how many ?
71. $6 \times 7 =$ how many ?
72. $6 \times 10 =$ how many ?
73. $6 \times 9 =$ how many ?
74. $6 \times 12 =$ how many ?
75. $6 \times 11 =$ how many ?
76. $7 \times 6 =$ how many ?
77. $7 \times 8 =$ how many ?
78. $7 \times 7 =$ how many ?
79. $7 \times 10 =$ how many ?
80. $7 \times 9 =$ how many ?
81. $7 \times 12 =$ how many ?
82. $7 \times 11 =$ how many ?

83. What will 5 barrels of flour cost, at 6 dollars a barrel ?
84. What will 5 bushels of potatoes cost, at 5 dimes a bushel ?

85. What will 6 primers cost, at 6 cents each ?
86. What will 5 barrels of fish cost, at 7 dollars a barrel ?
87. What will 6 pounds of mutton cost, at 7 cents a pound ?
88. What will 5 barrels of sugar cost, at 12 dollars a barrel ?
89. What will 6 pounds of sturgeon cost, at 10 cents a pound ?
90. What will 6 pounds of almonds cost, at 12 cents a pound ?
91. What will 5 barrels of pork cost, at $10 a barrel ?
92. What will 6 pounds of candles cost, at 9 cents a pound ?
93. What will 5 coats cost, at 9 dollars each ?
94. What will 6 handkerchiefs cost, at 11 cents each ?
95. What will 6 inkstands cost, at 8 cents each ?
96. What will 7 lamps cost, at 9 dimes each ?
97. What will 7 plows cost, at 8 dollars each ?
98. What will 7 boxes of caps cost, at 10 cents a box ?
99. What will 7 quires of paper cost, at 12 cents a quire ?
100. What will 7 letter-folders cost, at 11 cents each ?
101.   Eight   times   8   are   how   many ?
102.   Eight   times   10   are   how   many ?
103.   Nine   times   8   are   how   many ?
104.   Eight   times   7   are   how   many ?
105.   Nine   times   9   are   how   many ?
106.   Eight   times   9   are   how   many ?
107.   Nine   times   11   are   how   many ?
108.   Eight   times   12   are   how   many ?
109.   Nine   times   10   are   how   many ?
110.   Eight   times   11   are   how   many ?
111.   Nine   times   12   are   how   many ?
112. What will 9 bunches of roses cost, at 9 cents a bunch ?
113. What will 8 pen-knives cost, at 12 cents each ?

114. What will 9 bunches of grapes cost, at 12 cents a bunch?

115. What will 11 yards of calico cost, at 11 cents a yard?

116. What will 10 balls of cotton cost, at 12 cents a ball?

117. What will 11 pounds of ginger cost, at 12 cents a pound?

118. What will 10 blocks of tape cost, at 8 cents a block?

119. What will 12 yards of cloth cost, at 12 dimes a yard?

120. What will 13 pairs of boots cost, at $4 a pair?

LESSON IX.

ADDITION, SUBTRACTION, AND MULTIPLICATION COMBINED.

1. At 7 cents each, what will 9 pine-apples cost?

2. If the postage on 1 letter is 3 cents, what will be the postage on 8 letters?

3. If it require 8 yards of calico to make 1 dress, how many yards will it require to make 7 dresses?

4. If John obtain 2 credit-marks in one day, how many will he have in 15 days?

5. A man hired a horse to ride 12 miles, at the rate of 5 cents a mile; how much must he pay?

6. Margaret's cloak contains 7 yards of merino, worth 9 dimes a yard; what is the value of her cloak?

7. If a stage-coach go 9 miles in an hour, how far will it go in 7 hours?

8. At 8 dollars a week, how much will 5 weeks' board amount to?

9. If the fare by railroad from Albany to Boston is 5 dollars for 1 person; how much will it be for a family of 9 persons?

10. Helen had 8 rose bushes, and there were 7 roses on each; how many roses had she in all?

11. At 3 dimes a gallon, what will 15 gallons of molasses cost?

NOTE.—*One* dime = 10 cents.

12. There are 10 rows of trees in an orchard, and 12 trees in each row; how many trees are there in the orchard?

13. A traveler meeting 13 beggars, gave to each of them 3 dimes; how many dimes did he give to all of them?

14. A woman bought 11 yards of cloth, and paid for it with butter, giving 9 pounds for a yard; how many pounds of butter did it take to pay for the cloth? How much did the cloth cost, provided the butter was worth 10 cents a pound?

15. In a certain corn field there are 24 rows, and 30 hills in each row; how many hills in the field?

16. What will 40 steel pens cost, at 2 cents each?

17. What will 8 pairs of snuffers cost, at 3 dimes a pair?

18. When 2 dimes are paid for 1 duck, what will be the cost of 8 ducks? of 10 ducks? of 12 ducks?

19. When hay is worth 8 dollars a ton, what is the value of 2 tons? of 4 tons? of 3 tons? of 7 tons? of 5 tons? of 10 tons? of 12 tons? of 14 tons?

20. At 2 dimes each, how many cents will 4 books cost? 6 books? 10 books? 12 books? 11 books? 7 books? 16 books? 13 books? 14 books?

21. If 5 cents will buy one primer, what will be the cost of 4 primers? of 6? of 9? of 8? of 10?

22. 6 plates, at 5 dimes each, will cost how much?

23. At 5 dimes each, how much will 4 handkerchiefs cost? 6? 8? 10? 12? 14? 11? 16?

24. At 6 dimes each, how many cents will 2 geese cost? 4? 5? 8? 10? 12? 9? 7?

25. At 12 cents each, how much will 3 candlesticks cost? How much will 6? 5? 8? 9? 10? 7?

26. If I pay 5 cents for riding 1 mile, how much must I pay for riding 7 miles? 8 miles? 9? 6? 10? 12?

27. At 7 cents a yard, how much will 5 yards of ribbon cost? 6 yards? 8 yards? 9 yards? 10 yards? 12 yards?

28. If a tooth-brush cost 18 cents, how much will 4 cost?

29. 9 turkeys will cost how much, at 8 dimes each?

30. At 14 cents a quire, how much will 2 quires of paper cost? 3 quires? 4 quires? 5 quires?

31. How much will 7 pictures cost, at 5 cents each? at 6 cents each? at 8 cents each? at 10 cts. each?

32. How much will 8 knives cost, at 6 dimes each? at 10 dimes each?

33. At 10 dimes each, how much will 4 caps cost? 5? 6? 8? 9? 12? 14? 17? 19? 21? 25?

34. At 40 cents a day, how much will 2 days' work amount to? 5 days' work?

35. If one paper of candy cost 6 cents, how much will 3 papers cost? 5 papers? 8 papers? 12 papers?

36. At 7 dollars a hundred feet, how much will 4 hundred feet of cedar boards cost? 9 hundred feet? 10 hundred feet?

37. If 1 bushel of wheat cost 60 cents, how much will 6 bushels cost? 4 bushels? 5 bushels?

38. How much will 8 muffs cost, at 12 dollars each?

39. How much will 19 lead pencils cost, at 5 cents each?

40. How much will 11 boxes of cheese cost, at 4 dollars a box? at 5 dollars a box? at 8 dollars a box?

41. How much will 12 barrels of pork cost, at 5 dollars a barrel? at 8 dollars? at 9 dollars? at 10 dollars?

42. How much will 9 tons of hay cost, at 13 dollars a ton?

43. James is 9 years old, and his father is 4 times as old; how old is his father?

44. Jane's frock contains 7 yards of silk, worth 8 dimes a yard; what was the value of the silk? Provided the making cost 2 dollars, how much was the cost of her dress?

45. If a barrel of flour will serve 12 men 8 days, how long will it serve 1 man?

46. If I earn 12 dollars in a month, and spend 8, how much will I have at the end of 12 months?

47. If I earn 12 dollars a month, and pay 25 cents a week for washing, and 2 dollars a week for board, how much will I have at the end of 40 weeks (10 months)?

48. If I buy 9 tons of hay, at 12 dollars a ton, and sell 6 tons, at 15 dollars a ton, and the 3 remaining tons, at 10 dollars a ton; how much shall I gain?

49. Bought 11 yards of broadcloth, at 4 dollars a yard, but, being damaged, I lost 18 dollars by the sale of it; how much did I receive for it?

50. If I buy 12 barrels of pork, at 8 dollars a barrel, and sell it all for 108 dollars, how much shall I gain by so doing?

51. A man bought a horse for 80 dollars, paid 2 dollars a week for his keeping, and received 4 dollars a week for his work; at the expiration of 10 weeks he sold him for 70 dollars; how much did he gain?

52. For how much must I sell 4 barrels of wheat which cost me 8 dollars a barrel, to gain 8 dollars?

53. What is the cost of 9 cows, at 25 dollars each?

54. Provided a hunter should kill 5 pigeons, and wound 4 at every shot, how many would he kill and wound by shooting 8 times?

55. If a man travel 20 miles in a day, how many miles will he travel in 6 days?

56. How much will 8 months' wages amount to, at 18 dollars a month?

57. If 10 men eat 18 pounds of butter in 1 week, how long would it last one man?

58. If 80 dollars will pay for 4 dinners for 20 men, how many dinners would it buy for 1 man?

59. Bought 3 yards of cloth for a coat, at 7 dollars a yard, the buttons and cord cost 2 dollars, buckram and wadding, 1 dollar, paid for making it 6 dollars; for how much must I sell it to gain 5 dollars?

60. If 17 men can do a piece of work in 9 days, how many days would it take 1 man to perform the same work?

61. Two men start from the same place, and travel in opposite directions, one at the rate of 7 miles an hour, the other, 9 miles an hour; how far apart will they be in 2 hours?

62. Two men start from the same place, and travel the same way, one at the rate of 3 miles an hour, the other, 8 miles an hour; how far apart will they be at the end of 8 hours?

63. Two men are 50 miles apart, and approach each other, one at the rate of 2 miles an hour, the other, 3 miles an hour; how far apart will they be at the end of 5 hours?

64. If 1 orange is worth 4 apples, how many apples must be given for 13 oranges?

65. A boy earned 80 cents a day, and paid 50 cents a day for his board and washing; how much had he left at the expiration of 6 days?

66. Jane bought 4 yards of silk, at 2 dollars a yard, 3 shawls, at 10 dollars each, and some delaine for 10 dollars; she paid 5 ten-dollar bills; how much ought she to receive back?

67. Mary bought 5 yards of silk, at 8 dimes a yard, and 8 yards of linen, at 9 dimes a yard; how many yards did she buy, and how much did all cost?

68. In a certain school there are 12 girls, and 3 times as .many boys, less 8; how many boys in the school, and how many boys and girls together?

59. John has 7 books, and Mary has 4 times as many, less 18; how many has Mary, and how many have both?

70. Albert has 9 marbles, Aaron 3 times as many, less 7, and Amos has twice as many as both, less 8; how many has each, and how many have they together?

71. Perry worked for Elisha 4 days, at 6 dimes a day; Elisha gave him 7 bushels of corn, at 3 dimes a bushel; how much was then due Perry?

72. A merchant bought 25 pounds of sugar for 125 cents, and sold 15 pounds of it, at 6 cents a pound, and the remaining 10 pounds, at 4 cents a pound; how much did he gain by so doing?

73. If the interest on 1 dollar for a year is 6 cents, how much is the interest on 13 dollars for the same time?

74. What will 27 pounds of beef cost, at 4 cents a pound?

75. When beef is 5 cents a pound, and pork 9 cents, how much more will 9 pounds of pork cost than 9 pounds of beef?

76. Mary bought 35 quarts of milk, and on her way home she spilled 4 times 2 quarts, less 3 quarts; how many quarts had she remaining?

77. James has 9 walnuts, John twice as many, less 8, and Joseph twice as many as James and John, + 7; how many has each, and how many have all?

DIVISION.

Division is the process of finding how many times one number contains another.   •

Also, of finding one of the equal parts of a number.

The Terms of Division are the Dividend, Divisor, Quotient, and Remainder.

1. *The Dividend* is the number to be divided;

2. *The Divisor* is the number to divide by;

8. *The Quotient* is the number of times the dividend contains the divisor, or the value of the required part of the dividend;

4. *The Remainder* is the part of the dividend which is left, when the dividend does not contain the divisor an exact number of times. The remainder is always less than the divisor.

LESSON X.

I.   8   are   how   many   times   2 ?

ANALYSIS.—8 are as many times 2 as 2 is contained times in 8, which are 4 times.

2.	6	are	how	many	times	2 ?
3.	4	are	how	many	times	2 ?
4.	10	are	how	many	times	2 ?
5.	12	are	how	many	times	2 ?
6.	14	are	how	many	times	2 ?
7.	6	are	how	many	times	3 ?
8.	9	are	how	many	times	3 ?
9.	12	are	how	many	times	3 ?
10.	15	are	how	many	times	3 ?
11.	18	are	how	many	times	3 ?
12.	21	are	how	many	times	3 ?
13.	24	are	how	many	times	3 ?
14.	16	are	how .	many	times	2 ?
15.	18	are	how	many	times	2 ?

16. 20 are how many times 2 ?
17. 22 are how many times 2 ?
18. 24 are how many times 2 ?
19. 26 are how many times 2 ?
20. • 28 are how many times 2 ?

21. At 2 cents each, how many apples can you buy for 4 cents ?

ANALYSIS 1ST.—If for 2 cents I can buy 1 apple, for 4 cents I can buy as many apples as 2 cents is contained times in 4 cents, which are 2 times. Therefore, at 2 cents each, for 4 cents I can buy 2 apples.

ANALYSIS 2D.—If 2 cents will buy 1 apple, 4 cents will buy as many apples as 2 is contained times in 4, which is 2.

ANALYSIS 3D.—If for 2 cents I can buy 1 apple, for 1 cent I can buy one-half of an apple ; and for 4 cents, 4 times one-half, which are 4 halves, or 2 apples.

22. At 2 cents each, how many oranges can I buy for 6 cents ?
23. At 2 cents each, how many peaches can be bought for 8 cents ?
24. At 3 dimes a yard, how many yards of calico can be bought for 12 dimes ?
25. At 3 cents each, how many lemons can be bought for 9 cents ?
26. At 2 cents a yard, how many yards of tape can be bought for 10 cents ?
27. At 2 dimes a bushel, how many bushels of apples may be had for 12 dimes ?
28. How many pounds of ginger, at 2 dimes a pound, may be had for 14 dimes ?
29. How many baskets of strawberries, at 3 cents a basket, can be had for 15 cents ?
30. For 16 dollars, how many yards of cloth can be had, at 2 dollars a yard ?
31. For 18 apples, how many oranges can be bought, at the rate of 2 apples for 1 orange ?
32. How many primers, at 2 cents each, can be bought for 24 cents ?

33. How many barrels of flour, at 5 dollars a barrel, can be bought for 20 dollars?

34. For 22 dollars, how many sheep may be bought, at 2 dollars each?

35. How many melons may be had for 18 dimes, at 3 dimes each?

36. At 3 cents each, how many tops will 6 cents buy?

37. At 3 dimes a peck, how many beans will $2.10 buy?

38. At 3 cents a mile, how far can I ride for 24 cents?

39. At 4 dimes a bushel, how much rye will $1.20 buy?

40. How many books, at 4 dimes each, can be bought for 20 dimes?

The *Sign of Division* is ÷, and when placed between two numbers, it denotes that the one on the left is to be divided by the one on the right. Thus, $6 ÷ 3 = 2$, or $\frac{6}{3} = 2$ denotes that 6 is to be divided by 3, and that the quotient equals 2; and is read, 6 *divided by* 3 equals 2.

41. $8 ÷ 4 =$ how many?

42. $12 ÷ 4 =$ how many?

43. $16 ÷ 4 =$ how many?

44. $10 ÷ 5 =$ how many?

45. $15 ÷ 5 =$ how many?

46. $20 ÷ 5 =$ how many?

47. $28 ÷ 7 =$ how many?

48. $32 ÷ 4 =$ how many?

49. $30 ÷ 5 =$ how many?

50. $35 ÷ 5 =$ how many?

51. $36 ÷ 4 =$ how many?

52. $40 ÷ 5 =$ how many?

53. $44 ÷ 4 =$ how many?

54. $30 ÷ 6 =$ how many?

55. 48 contains 8 how many times?

ANALYSIS.—48 contains 8, 6 times [because 6 times 8 are 48].

56. 24 contains 8 how many times? 4? 12? 3?

57. 36 contains 9 how many times? 6? 3? 2?

58. 54 contains 2 how many times? 3? 9?

59. 75 contains 3 how many times? 5? 15?

60. 68 contains 2 how many times? 4?

61. At 5 dimes each, how many turkeys can be had for 25 dimes?

62. If the wages of 1 day is 4 dimes, what will be the wages for 9 days?

63. How many days will a man be required to work for 12 dimes, if he receive 4 dimes a day?

64. If a boy spends 5 cents a day, how many days will it take him to spend 15 cents?

65. A boy had 20 marbles, and divided them equally among his 5 brothers; how many did each receive?

66. A boy divided 28 cents equally among 4 poor women; how many cents did each receive?

67. A farmer gave 4 of his laborers 32 bushels of corn; how many bushels did each receive?

68. If 5 quarts of ale cost 30 cents, what will 1 cost?

69. At 5 cents a yard, how many yards of ribbon may be had for 35 cents? how many for 50 cents?

70. How many pine-apples, at 8 cents each, can be obtained for 40 cents? for 56 cents?

71. If a man travel 45 miles in 9 hours, how many miles does he travel in 1 hour?

72. If a man travel 5 miles in an hour, how many hours will it take him to travel 40 miles?

73. How many yards of cloth, at 4 dollars a yard, can you buy for 32 dollars?

74. In a certain orchard are 48 trees standing in rows, and 6 trees in each row; how many rows are in the orchard?

75. For 56 dollars, how many barrels of pork can be bought, at 8 dollars a barrel?

NOTE.—See Methods of Teaching and Key to Intellectual Arithmetic, page 28.

76. If a man can travel 6 miles in an hour, in how many hours can he travel 42 miles?

77. How many yards of cloth, at 4 dollars a yard, can you buy for 36 dollars?

78. A butcher gave 39 dollars for sheep, at the rate of 3 dollars each; bow many sheep did he buy?

79. 45 dollars were given for 9 barrels of flour; how much was it a barrel?

80. How long would it take to travel 72 miles, at the rate of 3 miles an hour?

LESSON XI.

1. 20 are how many times 2? 4? 6?
2. 22 are how many times 11?
3. 24 are how many times 3? 4? 2?
4. 25 are how many times 5?
5. 28 are how many times 2? 7?
6. 30 are how many times 2? 3? 5?
7. 32 are how many times 2? 4? 16? 8?
8. 34 are how many times 17?
9. 40 are how many times 2? 4? 5? 8?
10. 44 are how many times 2? 11?
11. 46 are how many times 23?
12. 48 are how many times 2? 3? 4? 6?
13. 50 are how many times 2? 10?
14. 56 are how many times 2? 7?
15. 57 are how many times 3?
16. 60 are how many times 2? 3? 4? 5? 6?
17. 64 are how many times 2? 4? 8?
18. 66 are how many times 2? 3? 6?
19. 68 are how many times 2? 4?
20. 70 are how many times 10? 2?
21. 72 are how many times 2? 4? 6? 8?
22. 5 are how many times 2, and how many remaining?

REMARK.—Whenever there is a remainder, it may be mentioned simply as a remainder.

23. 7 are how many times 2?
24. 17 are how many times 4? 2? 5?
25. 18 are how many times 6? 4? 2?
26. 34 are how many times 4? 6? 5? 2?
27. 25 are how many times 5? 4? 2? 3?
28. 16 are how many times 9? 4? 8? 7?
29. 32 are how many times 7? 5? 6?
30. 63 are how many times 9? 4? 5? 6?
31. 74 are how many times 2? 4? 6? 7?

32. 80 are how many times 2? 3? 4? 5? 6? 7? 8?
33. 84 are how many times 2? 3? 4? 5? 6? 7? 8?
34. 15 are how many times 4? 6? 7? 8?
35. 29 are how many times 2? 3? 4? 5? 6? 7?
36. 90 are how many times 2? 4? 6? 8? 9? 11?
37. 144 are how many times 2? 4? 6? 8? 12?

LESSON XII.

1. At 2 cents each, how many lemons can I buy for 14 cents ?

  ANALYSIS 1ST.—If 1 lemon cost 2 cents, for 14 cents I can buy as many lemons as 2 is contained times in 14, which are 7 times.

  ANALYSIS 2D.—If for 2 cents 1 can buy 1 lemon, for 1 cent I can buy 1 half of a lemon, and for 14 cents, 14 times 1 half, which are 14 halves, or 7 lemons.

2. How many boxes of cheese, at 4 dollars a box, may be had for 12 dollars ?

3. If one hundred pounds of hay cost 3 dollars, how many hundred may be had for 15 dollars ?

4. If one barrel of flour support 20 persons one week, how many persons will it support 4 weeks ?

5. If 1 man can ride 1 mile for 4 cents, how far can 2 men ride for 80 cents ?

6. If 10 men accomplish a certain piece of work in 2 days, how long will it take 1 man to do the same ?

7. If 3 yards of cloth make 1 coat, how many coats will 18 yards make ?

8. If I receive 12 dollars interest in one year, in how many years will I receive 144 dollars interest ?

9. A man traveled 7 miles in 1 hour, at the same rate how long would it take him to travel 63 miles ?

10. If 1 cow cost 13 dollars, how many cows may be had for 65 dollars ?

11. How many pens can you buy for 27 cents, if 1 pen cost 3 cents ?

12. If 8 apples are worth 40 chestnuts, how many chestnuts is 1 apple worth?

13. How many cents is 1 lemon worth, if 12 lemons are worth 48 cents?

14. How much will 1 cord of wood cost, if 20 cords cost 40 dollars?

15. If 1 pound of beef cost 7 cents, how much will 19 pounds cost?

16. For 147 cents, how many pounds of sugar can be bought, at 7 cents a pound?

17. If 9 yards of cloth cost 53 dollars, for how much must it be sold a yard to gain 10 dollars?

18. If 7 barrels of flour cost 38 dollars, and were sold at 7 dollars a barrel, what was the gain?

19. How many peaches, at 4 cents each, may be bought for 96 cents?

20. How many yards of cloth, at 4 dollars a yard can be bought for 116 dollars?

21. How many oranges, at 3 cents each, must be given for 18 lemons worth 4 cents each?

22. If 15 sheep cost 75 dollars, what will 1 sheep cost?

23. Will 4 barrels of wheat flour at 9 dollars a barrel, cost more than 12 barrels of corn at 4 dollars a barrel? how much more?

24. How many barrels of beef, at 3 dollars a barrel, can be bought for 54 dollars?

25. How many pounds of fish, at 5 cents a pound, may be had for 95 cents?

26. At 7 cents a pound, how many pounds of lead may be had for 84 cents?

27. How long will it require to travel 105 miles, at the rate of 5 miles an hour?

28. A person divided 72 peaches equally among 6 boys; how many did each receive?

29. 148 marbles were divided equally among some boys; how many boys were there, provided each boy received 2 marbles?

30. How many pounds of butter, at 14 cents a pound, can be bought for 28 apples, at 3 cents each?

31. At 7 cents a bottle, how many bottles of ink can you buy for 14 sheets of paper, at 2 cents a sheet?

32. In how many days can 1 man do as much as 7 men in 8 days?

33. In how many days can 2 men do as much work as 6 men in 3 days?

> ANALYSIS.—If 6 men can do a certain amount of work in 3 days, 1 man can do the same work in 6 times 3 days, which are 18 days; and 2 men can do it in one-half of 18 days, which is 9 days.

34. In how many days can 4 men earn as much as 8 men in 6 days?

35. In how many days can 15 men earn as much as 3 men in 25 days?

36. In how many months will 6 horses eat as much as 18 horses in 5 months?

37. How many men can in 7 days earn as much as 28 men in 4 days?

38. In 10 days 6 men will earn as much as how many men in 5 days?

39. How many yards of cloth, at 4 dollars a yard, may be had for 4 sets of chairs, at 12 dollars a set?

40. A farmer gave 13 barrels of flour, worth 4 dollars a barrel, for 26 yards of cloth; how much was the cloth a yard?

41. The dividend is 108, and the divisor is 12; required the quotient.

42. The dividend is 96, and the quotient is 4; what is the divisor?

43. The quotient is 7, and the divisor is 9; what is the dividend?

LESSON XIII.

THE FOUR FUNDAMENTAL RULES COMBINED.

1. 4 times 6 are how many times 2?

ANALYSIS.—4 times 6 are 24. 24 are 12 times 2. Therefore, 4 times 6 are 12 times 2.

2. 4 times 9 are how many times 3?
3. 4 times 8 are how many times 2?
4. 4 times 10 are how many times 5?
5. 4 times 12 are how many times 6?
6. 4 times 14 are how many times 7?
7. 5 times 9 are how many times 15?
8. 5 times 8 are how many times 4?
9. 5 times 12 are how many times 15?   6?
10. 6 times 7 are how many times 2?
11. 6 times 8 are how many times 12?   3?
12. 4 times 6 are how many times 8?
13. 7 times 15 are how many times 5?
14. 12 times 7 are how many times 21?
15. 8 times 7 are how many times 4?
16. How many times 12 are 9 times 4?
17. How many times 20 are 5 times 4?
18. How many times 9 are 3 times 21, +9?
19. How many times 5 are 7 times 15, +10−5?
20. How many times 9 are 3 times 36, −2+11?
21. How many times 12 are 9 times 4, +24−12?
22. How many times 21 are 9 times 14, +42?
23. How many times 7 are 3 times 14, +21−14?
24. How many times 5 are 8 times 10, +5−15?
25. How many times 5 are 10 times 6, +15+5?
26. How many times 5 are 6 times 15, +10+15?
27. 10 times 4, +2, are how many times 7?   2?
28. 8 times 9, −2, are how many times 5?
29. 12 times 8, −8, are how many times 2?
30. 26 times 11, −6, are how many times 28?
31. 7 times 8, +4, are how many times 12?

LESSON XIV.

QUESTIONS COMBINING MULTIPLICATION AND DIVISION.

1. If 2 apples cost 4 cents, how many cents will 3 apples cost ?

> ANALYSIS.—If 2 apples cost 4 cents, 1 apple will cost 1 half of 4 cents, which is 2 cents. If 1 apple costs 2 cents, 3 apples will cost 3 times 2 cents, which are 6 cents.

2. If 2 pears cost 16 cts., how much will 5 pears cost ?
3. If 4 quinces cost 12 cents, how much will 3 quinces cost ?
4. If 6 oranges cost 18 cents, how much will 9 oranges cost ?
5. If 7 peaches cost 21 cents, how much will 9 peaches cost ?
6. If 4 lemons cost 24 cents, how much will 7 lemons cost ?
7. If 3 yards of tape cost 18 cents, how much will 6 yards cost ?
8. If 7 hair-brushes cost 28 dimes, how many cents will 6 hair-brushes cost ?
9. If 9 yards of muslin cost 108 cents, how much will 7 yards cost ?
10. If 11 books cost 44 dimes, how many cents will 7 books cost ?
11. If 12 ink-stands cost 96 cents, what will 2 cost ?
12. If 10 lead-pencils cost 30 cents, how much will be the cost of 7 ? of 9 ? of 2 ? of 15 ? of 12 ?
13. How much will 13 yards of silk cost, if 5 yards cost 45 dimes ?
14. If a man travel 15 miles in 3 hours, how far, at this rate, can he travel in 9 hours ? 5 hours ? 7 hours ?
15. If the cartage of a load of plaster 20 miles cost 4 dollars, how far could it be carried for 12 dollars ?
16. How many pairs of pantaloons can be cut out of 21 yards of cloth, if 3 pairs can be cut out of 9 yards of the same kind of cloth ?

17. How much will 30 pounds of sugar cost, if 6 pounds cost 42 cents?
18. How much will 18 pounds of veal cost, if 6 pounds cost 42 cents?
19. How much will 75 pounds of pork cost, if 9 pounds cost 75 cents?
20. How much will 20 weeks' board amount to, if 7 weeks' board cost 35 dollars?
21. How much will be the wages for one year, if 4 months' wages amount to 48 dollars?
22. How much will be the cost of 25 bushels of apples, if 13 bushels cost 260 cents?
23. How much will 14 pounds of cheese cost, if 6 pounds cost 54 cents?
24. What cost 36 quarts of milk, at 35 cts. for 7 qt.?
25. If 4 men can do a piece of work in 12 days, in how many days can 3 men do the same work?
26. I gave 72 dollars for a quantity of cotton, and sold it for 12 yards of cloth; how much did the cloth cost me a yard?
27. Gave 15 pounds of sugar for 5 pounds of butter; how much did the butter cost a pound, provided 8 pounds of sugar were worth 56 cents?
28. If 4 chestnuts are worth 8 walnuts, how many walnuts are 27 chestnuts worth?
29. If 7 yards of broadcloth are worth 35 dollars, how many boxes of butter, at 3 dollars a box, would 9 yards of this cloth buy?
30. A man bought 4 barrels of flour for 20 dollars, and gave 3 of them for cider, at 3 dollars a barrel; how many barrels of cider did he get?
31. A man bought 14 barrels of cider for 56 dollars, and gave 5 barrels of it for cloth, at 2 dollars a yard; how many yards did he receive?
32. A merchant having 15 yards of cloth worth 75 dollars, gave 10 of them for pork, worth 10 dollars a barrel; how many barrels did he receive?

33. When 9 bushels of rye were worth 45 dimes, 12 bushels were given for 15 yards of cloth; how much did the cloth cost a yard?

34. If 35 yards of cloth cost 140 dollars, how much will 95 yards of the same cloth cost?

35. Two boys are 32 rods apart, and both running in the same direction, the hindermost boy gains on the other 4 rods each minute; in how many minutes will he overtake him?

36. How many boxes will be required to contain 56 bushels, provided each box contains 8 bushels?

37. How many barrels of onions, at 3 dollars a barrel, must be given for 21 boxes of raisins, at 2 dollars a box?

38. A farmer bought 9 yards of cloth, at 4 dollars a yard, and paid for it with cider, at 3 dollars a barrel; how many barrels did it take?

39. How long would it take a man to save 24 dollars, if he save 2 dollars a week?

40. A farmer hired a laborer and agreed to give him 6 dollars for every 3 days' work; how much did he receive a week, allowing 6 working days in a week? how much a month, allowing 4 weeks to the month?

41. If 4 oranges are worth 12 cents, how many oranges must be given for 6 pine-apples, worth 12 cents each?

42. How many yards of cloth, at $2 a yard, can be bought for 4 reams of paper, at 5 dollars a ream?

43. If 2 apples are worth 1 orange, and 2 oranges are worth 1 lemon; how many lemons can be bought for 48 apples?

44. If 7 men can do a certain job of work in 12 days, in how many days could 21 men do the same work?

45. Bought 5 firkins of butter for 35 dollars; for what must I sell it to gain 10 dollars; what is the gain on each firkin?

LESSON XV.

COMPOUND DENOMINATE NUMBERS.

TABLE OF UNITED STATES CURRENCY.

10 Mills (*m.*)	make	1 Cent,	marked	*ct.*
10 Cents	"	1 Dime,	"	*d.*
10 Dimes	"	1 Dollar,	"	$.
10 Dollars	"	1 Eagle,		*E.*

1. How many mills in 4 cents?

   ANALYSIS.—In 1 cent there are 10 mills, and in 4 cents there are 4 times 10 mills, or 40 mills.

2. How many mills in 3 cents? In 5 cents? In 8?

3. How many cents in 2 dimes? In 4 dimes? In 5 dimes? In 6 dimes? In 9 dimes? In 10 dimes?

4. How many dimes in $1? In $2? In $3? In $4? In $5? In $6?

5. How many dimes in 1 E. and $4? In 2 E. and $8?

6. How many cents in $1? In $2? In $3? In $4?

7. How many dollars in 80 dimes?

   ANALYSIS.—There are 10 dimes in $1; therefore, 1 tenth of the number of dimes equals the number of dollars. 1 tenth of 80 is 8. Therefore, in 80 dimes there are $8.

8. How many dimes in 60 cents? In 70 cents?

9. How many dollars in 200 cents? In 500 cents? In 800 cents? In 360 cents? In 705 cents?

10. If 3 yards of muslin cost 6 dimes, how many yards can be bought for $1?

LESSON XVI.

TABLE OF ENGLISH MONEY.

4 Farthings (*far.*)	make	1 Penny,	marked	*d.*
12 Pence	"	1 Shilling,	"	*s.*
20 Shillings	"	1 Pound,	"	£.

A Sovereign (*sov.*) is equal in value to £1.

1. How many shillings in £4 15 shillings?

> ANALYSIS 1ST.—There are 20 shillings in £1; therefore, 20 times the number of pounds equal the number of shillings. 20 times 4 are 80 shillings, +15, are 95 shillings.
>
> ANALYSIS 2D.— In £1 there are 20 shillings, and in £4 there are 4 times 20 shillings, which are 80 shillings. 80 shillings + 15 shillings, are 95 shillings.

2. How many pence in 1 shilling? In 4 shillings? In 3 shillings? In 7 shillings? In 9 shillings?
3. How many shillings in £1? In £2? In £3?
4. How many pence in £2 10 shillings 5 pence?
5. How many pounds in 60 shillings? In 80s.?
6. How many pounds in 480 pence? In 720d.?
7. At 4 shillings a bushel, how many pounds will 40 bushels of potatoes cost?
8. At 10 pence each, how many pounds will 48 pine-apples cost?
9. At 5 shillings a yard, how many yards of cloth can be bought for £2 15 shillings?

LESSON XVII.

TABLE OF TROY WEIGHT.

24 Grains (gr.)	make	1 Pennyweight,	marked	pwt.
20 Pennyweights	"	1 Ounce,	"	oz.
12 Ounces	"	1 Pound,	"	℔.

1. How many pennyweights in 240 grains?
2. How many pennyweights in 4 ounces? In 5 ounces? In 6 ounces?
3. How many ounces in 1 pound? In 3 pounds? In 5 pounds? In 8 pounds?
4. In 24 ounces how many pounds? In 48 ounces? In 36 ounces? In 60 ounces? In 84 ounces?
5. How many ounces in 20 pennyweights? In 40? In 60? In 70?

6. If 7 grains of gold cost 2 dimes and 8 cents, how much will 10 pennyweights cost?
7. How many pounds in 480 pennyweights?
8. How many grains in 2 oz., 2 pwt., and 2 gr.?

LESSON XVIII.

TABLE OF AVOIRDUPOIS WEIGHT.

16 Drams (*dr.*)	make 1 Ounce,	marked	*oz.*
16 Ounces	" 1 Pound,	"	*lb.*
25 Pounds	" 1 Quarter,	"	*qr.*
4 Quarters, or 100 lb.	" 1 Hundred-weight,	"	*cwt.*
20 Hundred-weight	". 1 Ton,	"	*T.*

1. How many drams in 2 oz.? In 4 oz.? In 10 oz.?
2. How many ounces in 2 lb.? In 4 lb.? In 8 lb.?
3. How many quarters in 100 pounds? In 400 pounds? In 600 pounds? In 1200 pounds?
4. How many pounds in 3 qr.? In 2 qr.? In 7 qr.?
5. How many quarters in 2 hundred-weight?
6. In 8 quarters how many hundred-weight?
7. What will 2 tons of iron cost, if 1 lb. cost 1 dime?
8. What will 40 tons of hay cost, at 2 dimes a qr.?

LESSON XIX.

TABLE OF LONG MEASURE.

12 Inches (*in.*)	make 1 Foot,	marked	*ft.*
3 Feet	" 1 Yard,	"	*yd.*
5½ Yards, or 16½ feet,	" 1 Rod,	"	*rd.*
40 Rods	" 1 Furlong,	"	*fur.*
8 Furlongs, or 320 rods,	" 1 Mile,	"	*mi.*
3 Miles	" 1 League,	"	*lea.*
69½ Miles	" 1 Degree,	"	*deg.*, or °
360 Degrees	" 1 Circle of the earth.		

3

1. How many inches in 1 ft. ?   In 2 ft. ?   In 4 ft. ?
In 5 ft. ?   In 10 ft. ?
2. How many feet in 2 yd. ?   In 6 yd. ?   In 5 yd. ?
3. How many yards in 2 rods ?   In 4 rd. ?   In 8 rd. ?
4. How many furlongs in 4 miles ?   In 6 mi. ?
5. How many inches in 4 yd. 2 ft. 10 in. ?
6. In 216 inches, how many yards ?   In 288 in. ?
7. How many rods in 1 mile ?
8. How many miles in 1760 yards ?
9. How many feet in 2 rods 3 yd. 2 ft. 11 in. ?
10. How many feet in 1 mile ?

LESSON XX.

TABLE OF CLOTH MEASURE.

In Cloth Measure, the *Yard* is the unit of measure, and is
divided into *Halves, Fourths, Eighths,* and *Sixteenths.* The following Table was formerly used :

2¼ Inches	make	1 Nail,	marked	*na.*
4 Nails	"	1 Quarter of a yard,	"	*qr.*
4 Quarters	"	1 Yard,	"	*yd.*
3 Quarters	"	1 Ell Flemish,	"	*E. Fl.*
5 Quarters	"	1 Ell English,	"	*E. E.*
6 Quarters	"	1 Ell French,	"	*E. Fr.*

1. In 4 yd. 3 qr., how many quarters ?
2. In 7 yd. 2 qr., how many Ells French ?
3. In 3 yd. 3 qr., how many Ells Flemish ?

LESSON XXI.

TABLE OF LAND MEASURE.

144	Square inches (*sq. in.*)	make	1 Square foot,	*sq. ft.*
9	Square feet	"	1 Square yard,	*sq. yd.*
30¼	Square yards	"	1 Square rod, or pole,	*P.*
40	Square rods	"	1 Rood,	*R.*
4	Roods	"	1 Acre,	*A.*
640	Acres	"	1 Square mile,	*sq. m.*

1. How many square feet in 4 sq. yd.? In 8 sq. yd.?
2. How many poles in 4 roods? In 6 roods?
3. How many acres in 40 roods? In 160 roods?
4. How many square yards in 81 sq. ft.? In 108 sq. ft.?
5. How many square yards in 1 rood 10 rods?

LESSON XXII.

TABLE OF CUBIC MEASURE.

1728	Cubic inches (*cu. in.*) make	1 Cubic foot,	*cu. ft.*
27	Cubic feet "	1 Cubic yard,	*cu. yd.*
24¾	Cubic feet "	1 Perch of stone,	*pch.*
16	Cubic feet "	1 Cord foot,	*c. ft.*
8	Cord feet, or } "	1 Cord of wood,	*C.*
128	Cubic feet }		

1. How many cubic feet in 4 cubic yards?
2. How many cubic feet in 4 perch of stone?
3. How many cords in 96 cord feet? In 72 c. ft.?
4. How many cords in 128 cu. ft.? In 384 cu. ft.?
5. How many cord feet in 4 cords? In 6 cords?

LESSON XXIII.

TABLE OF WINE MEASURE.

4	Gills (*gi.*) make	1 Pint,	marked	*pt.*
2	Pints "	1 Quart	"	*qt.*
4	Quarts "	1 Gallon,	"	*gal.*
42	Gallons "	1 Tierce,	"	*tier.*
31½	Gallons "	1 Barrel,	"	*bar.*
2	Barrels, or 63 Gallons, "	1 Hogshead,	"	*hhd.*
2	Hogsheads "	1 Pipe,	"	*pi.*
2	Pipes "	1 Tun,	"	*tun.*

NOTE.—Barrels, tierces, hogsheads, pipes, and tuns are not measures, but vessels of variable capacity, which are gauged, the contents being sold by the gallon.

1. How many gills in 3 pints ?   In 4 pints ?
2. In 3 qt., how many gills ?
3. In 12 gallons, how many pints ?
4. What will 5 gall. of rum cost, if 4 gi. cost 5 cents?
5. How many pints in 2 pipes ?
6. A merchant bought a hogshead of molasses for 20 dollars, and sold it at the rate of 15 cents for 3 pints; how much did he gain by the bargain ?
7. What will 1 gall. of wine cost, at 21 cents for 7 gi.?
8. In 4 quarts and 2 pints, how many gills ?

LESSON XXIV.

TABLE OF DRY MEASURE.

2 Pints	make	1 Quart,	marked	*qt.*
8 Quarts	"	1 Peck,	"	*pk.*
4 Pecks	"	1 Bushel,	",	*bu.*

1. In 1 peck, how many pints ?
2. 2 pecks will fill how many pint measures ?
3. In 3 pecks and 3 quarts, how many pints ?
4. In 1 bushel and 3 pecks, how many quarts ?
5. In 1 bushel, how many quarts ? how many pints ?
6. If 8 pints of nuts cost 24 cents, what will 3 pecks cost at the same rate ?
7. A market woman bought 4 quarts of strawberries for 29 cents, and sold them at 5 cents a pint; how much did she gain ?
8. A person sold 2 bushels and 1 peck (or 144 pints) of currants, at 2 cents a pint, and in payment received 1 bushel (or 64 pints) of gooseberries, at 4 cents a pint; how much remains due ?
9. What will 5 quarts of wheat cost, if 1 bushel cost 128 cents ?

LESSON XXV.

TABLE OF TIME.

60	Seconds (*sec.*)	make	1 Minute,	marked *m.*
60	Minutes	"	1 Hour,	" *hr.*
24	Hours	"	1 Day,	" *d.*
7	Days	"	1 Week,	" *w.*
4	Weeks	"	1 Month,	" *mo.*
12	Calendar Months	"	1 Year,	" *yr.*
52	Weeks	"	1 Year,	" *yr.*
365	Days	"	1 Common Year,	" *yr.*
366	Days	"	1 Leap Year,	" *yr.*
100	Years	"	1 Century,	" *C.*

The following Table exhibits the divisions of the year, the names of the months, and the number of days in each :—

Winter.	1st month,	January,	has	31 days.
	2d month,	February,	has	28, in leap year 29.
Sprng.	3d month,	March,	has	31 days.
	4th month,	April,	has	30 days.
	5th month,	May,	has	31 days.
Summer.	6th month,	June,	has	30 days.
	7th month,	July,	has	31 days.
	8th month,	August,	has	31 days.
Autumn.	9th month,	September,	has	30 days.
	10th month,	October,	has	31 days.
	11th month,	November,	has	30 days.·
Winter.	12th month,	December,	has	31 days.

The following lines will help to remember the number of days in each month :

> "Thirty days hath September,
> April, June, and November;
> All the rest have thirty-one,
> Except February alone,
> Which hath but twenty-eight in fine,
> Till leap year gives it twenty-nine."

☞ *In our calculations on interest we shall reckon 30 days to the month, and 12 months to the year, although not strictly accurate.* [*See New Practical Arithmetic, page* 184.]

1. In 2 hours, how many seconds?
2. In 2 weeks and 5 days, how many days?
3. In 48 hours, how many days?
4. In 7200 seconds, how many hours?
5. How many hours in a week?
6. In 1 day 12 hours and 10 minutes, how many minutes?
7. How many hours in a month?
8. If a boy can do a certain piece of work in 40 minutes, how many hours would it take him to perform 12 times as much work?
9. If I can do a piece of work in 10 minutes, how many hours would it take to perform 12 times as much work?
10. How many days in 3 weeks and 5 days?

LESSON XXVI.

MISCELLANEOUS TABLE.

12 Units	make	1 Dozen.
12 Dozen	"	1 Gross.
12 Gross	"	1 Great Gross.
20 Units	"	1 Score.
24 Sheets of Paper	"	1 Quire.
20 Quires	"	1 Ream.
56 Pounds	"	1 Bushel of Corn or Rye.
60 Pounds	"	1 Bushel of Wheat.
196 Pounds	"	1 Barrel of Flour.
200 Pounds	"	{ 1 Barrel of Beef, Pork, or Fish.

1. What will 2 reams of paper cost, at 15 cents a quire?
2. How many sheets of paper in 1 ream?
3. How many years in "3 score years and 10"?
4. How many units in a gross?

LESSON XXVI.

METRICAL WEIGHTS AND MEASURES.

In France and several other countries of Europe, the division of Weights and Measures is on the decimal system. On this system the United States currency is founded, and its application to Weights and Measures is now permitted by Congress.

The METRE is the unit in measuring lengths, and is the basis of the system. It is one ten-millionth part of a quadrant of the earth's meridian, or that part of the earth's circumference between the Equator and the Pole. The Metre is nearly equal to 3 feet and $3\frac{3}{8}$ inches. 5 Metres = 1 Rod nearly.

The ARE is the unit for superficial measure. It equals the area of a square whose side is ten metres.

The STERE is the unit for wood and cubic measure. It is a cubic metre, and contains 100 litres.

The LITRE is the unit in measures of capacity, and is a cubic decimetre. The litre is equal to a little more than a quart wine measure.

The GRAM is the unit of weight, and is the weight of a cube of pure water at the temperature of melting ice, each edge of the cube being $\frac{1}{100}$th of a metre. The gram is equal to very nearly $15\frac{1}{2}$ grains Troy weight.

REMARK.—Each of these units is divided decimally, and larger units are formed by multiples of 10, 100, etc. The successive multiples are designated by the prefixes *deka*, *hecto*, *kilo*, *myria;* the parts by *deci*, *centi*, and *milli*, as shown by the following Tables.

The FRANC is the unit of value, and is a coin weighing five grains. It is equal to 18 cents 6 mills U. S. coin. The Dollar is the unit of value in U. S. currency.

Note.—For the Numeration and Notation of Integers and Decimals, see pages 171, 172.

MEASURES OF LENGTH.

10 Millimetres	make	1 Centimetre	=	.393685 of an Inch.
10 Centimetres	"	1 Decimetre	=	3.93685 Inches.
10 Decimetres	"	1 METRE	=	39.3685 "
10 Metres	"	1 Dekametre	=	393.685 "
10 Dekametres	"	1 Hectometre	=	3936.85 "
10 Hectometres	"	1 Kilometre	=	39368.5 "
10 Kilometres	"	1 Myriametre	=	393685 "

1. How many decimetres are in 1 metre? In 3 metres? In 8 metres? In 1 dekametre?
2. How many centimetres are in 1 metre?
3. How many millimetres are in 1 metre? In 1 decimetre? In 1 dekametre? In 1 hectometre? In 1 kilometre? In 1 myriametre?
4. Then how many metres are in one myriametre?
5. One metre is equal to how many inches of Long Measure? A myriametre to how many inches?

MEASURES OF WEIGHT.

DENOMINATION.	DECIMAL EXPRESSION.		COMPARATIVE WEIGHT.
10 Milligram make 1	.001	=	.015434 of a Grain (Troy weight).
10 Centigram make 1	.01	=	.15434 " "
10 Decigram make 1	.1	=	1.5434 Grains (Troy weight).
10 GRAM make 1	1	=	15.434 " "
10 Dekagram make 1	10	=	154.34 " "
10 Hectogram make 1	100	=	1543.4 " "
10 Kilogram	1000	=	15434 " "

MEASURES OF CAPACITY, ETC.

10 Millilitres	make	1 Centilitre	=	.61028 of a Cubic Inch.
10 Centilitres	"	1 Decilitre	=	6.1028 Cubic Inches.
10 Decilitres	"	1 LITRE	=	61.028 " "
10 Litres	"	1 Dekalitre	=	610.28 " "
10 Dekalitres	"	1 Hectolitre	=	6102.8 " "
10 Hectolitres	"	1 Kilolitre	=	61028. " "

100 Centiares   make   1 ARE. | 100 Ares · make   1 Hectare.

10 Centisteres make 1 Decistere.| 10 Decisteres make 1 STERE.
10 Steres make 1 Dekastere.

FRENCH CURRENCY.

10 Millimes	make	1 Centime	=	.001 of a dollar.
10 Centimes	"	1 Decime	=	.018 "
10 Decimes	"	1 Franc	=	.186 "
5 Franc piece, (silver)			=	.930 "
20 Franc piece, or Napoleon (gold)			=	3.72 dollars.

FRACTIONS.

LESSON XXVII.

1. John has 6 nuts, and Joel 1 half as many; how many has Joel?

   ANALYSIS.—If John has 6 nuts, and Joel 1 half as many, Joel must have 1 half of 6 nuts, which is 3 nuts.

2. Mary had 4 dresses, and Rachel 1 half as many; how many had Rachel?

3. Jacob is 8 years old, and John is 1 half as old; how old is John?

4. Moses had 2 marbles, and gave 1 half of them to his brother; how many had he left?

5. If you divide 6 apples equally between 2 boys, what part of them will each have?

6. What is 1 half of 6?

7. How many halves in 1?

8. If an orange cost 8 cents, and a peach 1 half as much; what was the cost of the peach?

9. James had 12 cakes, and John 1 half as many; how many had John?

10. If 3 apples cost 6 cents, what part of 6 cents will 1 apple cost? 2 apples? 3 apples?

11. What is 1 third of 6? 2 thirds of 6?

12. What is 1 half of 8? 10? 12? 14? 16? 18?

13. If 3 quarts of strawberries cost 18 cents, what will 1 quart cost? 2 quarts?

14. What is 1 third of 18? 1 half of 18?

15. If 4 pounds of raisins cost 8 dimes, what part of 8 dimes will 1 pound cost? 2 pounds? 3 pounds?

16. What is 1 fourth of 8? of 12? of 16? of 20?

17. What is 1 fifth of 15? of 10? of 20? of 30?

18. If 1 fifth of 15 is 3, what is 2 fifths of 15? 3 fifths? 4 fifths? 6 fifths? 8 fifths?

19. What is 1 sixth of 12?

20. What is 2 sixths of 12? 3 sixths? 4 sixths? 5 sixths?
21. What is 1 seventh of 21?
22. If 1 seventh of 21 is 3, what is 2 sevenths of 21? 3 sevenths? 4 sevenths? 5 sevenths?
23. If 1 pound of candy cost 12 cents, what part of a pound can you buy for 1 cent? for 2 cents? for 3 cents? for 5 cents? for 8 cents?
24. If a coat cost $20, and a pair of pantaloons 1 fourth as much, how much will the pantaloons cost?
25. If 7 barrels of cider cost $28, what part of $28 will 1 barrel cost? 4 barrels? 6 barrels? 5 barrels?
26. What is 1 seventh of $28? 2 sevenths of $28? 4 sevenths? 5 sevenths? 7 sevenths? 6 sevenths?
27. If 1 pound of cheese cost 6 cents, how much will 1 third of a pound cost? 2 thirds?
28. If 12 lemons cost 36 cents, what part of 36 cents will 1 lemon cost? 2? 4? 5? 8? 10? 9? 7?
29. What is 1 twelfth of 36? 2 twelfths of 36? 4 twelfths? 5 twelfths? 6 twelfths? 9 twelfths?
30. What do you understand by 1 third? 2 thirds?

ANSWER.—When a thing has been divided into three equal parts, one of these parts is called 1 *third*, and 2 of these parts are called 2 *thirds*.

31. What do you understand by 1 half?
32. What do you understand by 1 fourth? 2 fourths?
33. What do you understand by 1 fifth? 2 fifths?
34. How many thirds make a whole one?
35. How many fourths in 1?
36. What do you understand by 2 sixths? 4 sixths?
37. What do you understand by 3 sevenths? 2 sevenths? 4 sevenths? 5 sevenths?
38. How many sixths in 1?
39. How many ninths in 1?
40. How many eighths in 1?
41. How many sevenths in 1?
42. How many tenths in 1?
43. How many twentieths in 1?

44. What do you understand by 7 twelfths? 6 twelfths? 9 twelfths? 8 twelfths?

45. James had 9 marbles, and Jacob had 2 thirds as many; how many had Jacob?

ANALYSIS.—If James has 9 marbles, and Jacob 2 thirds as many, Jacob must have 2 thirds of 9 marbles. 1 third of 9 marbles is 3 marbles, and 2 thirds are 2 times 3 marbles, which are 6 marbles, Jacob's number.

46. Mary bought 12 needles, and Sarah bought 2 thirds as many; how many did Sarah buy?

47. Rachel has 12 primers, Mary 3 fourths as many, and Anthony 2 thirds as many as Mary; how many have Mary and Anthony respectively?

48. Albert is 15 years old, and Ebenezer is 4 fifths as old; how old is Ebenezer?

49. Augustus has 40 cents, and Augusta has 5 eighths as many; how many has she?

50. Morgan had 36 marbles, and gave 4 sixths of them to Martin; how many did he give to Martin, and how many had he left?

51. Moses had 24 fire-crackers, and Nathan has 7 sixths as many; how many has Nathan?

52. Mifflin had 45 cents, and Matthew had 5 ninths as many; how many had Matthew?

53. Dubois is 20 years old, and his father is 9 fifths as old; what is his father's age?

54. A farmer had 84 sheep, and a wolf killed 1 third of them; how many had he remaining?

55. In a certain school there are 12 girls, and 7 fourths as many boys; required the number of boys, and the number of boys and girls together.

56. In a recitation 36 questions were answered, and 1 ninth of them answered wrong; how many were correctly answered?

57. 4 fifths of all the words given out in a lesson were spelled correctly, and 8 were misspelled; of how many words did the lesson consist?

58. Montgomery bought 9 walnuts for 1 cent; what part of a cent did 1 cost? 2? 3? 6? 7? 9?

59. A colt was bought for $60, and sold for 7 fifths of what it cost; how much was the gain?

60. A. received $140 for 14 weeks' labor, and paid 1 fifth of it for board; how much did he save each week?

61. How many are 4 fifths of 75?

62. How many are 7 eighths of 24?

63. Mr. A.'s wife is 40 years old, and 9 eighths of her age equals his; what is his age?

64. What is 2 ninths of 36? 4 ninths? 3 fourths? 4 sixths? 5 sixths? 4 twelfths? 9 twelfths?

65. How many are 3 fourths of 48? 4 sixths? 5 eighths? 7 eighths? 6 eighths? 5 sixths? 2 thirds?

66. 3 ninths of 27 are how many? 4 ninths? 7 ninths? 8 ninths? 2 thirds?

67. 3 fourths of 24 are how many times 3?

ANALYSIS.—1 fourth of 24 is 6, and 3 fourths are 3 times 6, which are 18. 18 are 6 times 3.

68. 5 sevenths of 63 are how many times 3?

69. 3 eighths of 64 are how many times 6?

70. 9 thirds of 18 are how many times 3?

71. 4 fifths of 25 are how many times 2?

72. 6 ninths of 18 are how many times 6?

73. 7 ninths of 90 are how many times 2?

74. 4 thirds of 39 are how many times 2?

75. 15 seventeenths of 34 are how many times 6?

76. How many times 17 are 17 eighteenths of 36?

77. How many times 8 are 12 thirteenths of 26?

78. How many times 5 are 10 thirds of 36?

79. How many times 4 are 2 thirds of 27, −2?

80. How many times 6 are 3 halves of 48, +12?

81. Stephen having 40 apples, gave 3 fifths of them to one companion, and 3 eighths of them to another; how many had he remaining?

82. A. had $120; 1 third of it he spent for a watch,

1 fourth of it for a suit of clothes, and 3 tenths of it for a sleigh; how much had he remaining?

83. B., being asked the cost of his hat, replied, 2 thirds of $30 is 4 times its cost; required its cost.

84. 14 ninths of $27 is equal to 7 times the cost of a pair of boots; required the cost of the boots.

85. An individual, having $90 on interest, received 2 forty-fifths of the principal for the interest; how much interest did he receive?

86. The interest received on $360, was 1 eighteenth of the principal; how much was the interest?

87. B. is worth $2000, and 3 fourths of his fortune is 3 times A.'s; required A.'s fortune.

88. 3 eighths of the number of hours in a day, is 3 times the number of hours I work; how many hours do I work?

89. A pole, whose length is 16 feet, is in the air and water; and 3 fourths of the whole length, less 4 feet, equals the length in the air; required the length in the water.

90. 3 fifths of $2000, +$120, equals B.'s fortune; how much is B. worth?

91. The building of a house cost $560, and 4 sevenths of this, +$80, is 1 tenth of the cost of the farm on which it stands; required the cost of the farm.

92. 5 eighths of 72, +13, are how many times 2?

93. The interest on $960 for 5 years, was equal to 1 third of the principal; how much was the yearly interest?

94. What will 2 thirds of 12 pounds of coffee cost, at 13 cents a pound?

95. What will 3 fourths of a gallon of alcohol cost, at 9 cents a pint?

96. What will 1 sixteenth of a bushel of flax-seed cost, at 5 cents a pint?

97. How much will 7 fifteenths of 30 pine-apples cost, at 2 dimes each?

98. How much will 7 ninths of a hogshead of molasses cost, at 4 dimes a gallon ?

99. How many cents will 3 fifths of 100 oranges cost, at 1 half dime each ?

100. If 1 pennyweight of gold cost $1, how much will 2 fifths of an ounce cost ?

101. What will be the cost of 2 thirds of 36 pounds of butter, at 2 dimes a pound ?

102. 2 thirds of 24, +3 fourths of 16, are how many times 7 ?

103. 2 thirds of 30, +3 fifths of 40, are how many times 8 ?

104. 3 sevenths of 21, + 3 eighths of 40, are how many times 6 ?

105. How far can I walk in 3 eighths of a day, at the rate of 3 miles an hour ?

106. If Marcus earn 1 dime in an hour, how many cents can he earn in 3 eighths of a day ?

107. If a horse travel 10 miles in an hour, how many times 10 miles can he travel in 5 twelfths of a day ?

108. How much will 1 quart of gin cost, if 1 gill cost 15 cents ?

109. How many dollars will 4 sixths of a pound of gold cost, if 1 pennyweight cost 10 dimes ?

110. How many Eagles will 25 fourths of a gallon of brandy cost, at 1 half-dime a gill ?

LESSON XXVIII.

1. If 1 third of an orange cost 2 cents, what will 1 orange cost?

   ANALYSIS.—If 1 third of an orange cost 2 cents, 3 thirds, or 1 orange, will cost 3 times 2 cents, which are 6 cents.

2. If 1 half of a pound of raisins cost 8 cents, what will 1 pound cost?

3. Bought 1 third of a barrel of sugar for $3; how much will 2 thirds of a barrel cost, at the same rate?

4. If 1 third of a pound of pork cost 5 cents, how much will 2 pounds cost?

5. 2 is 1 third of what number?

   ANALYSIS.—If 1 third of some number is 2, 3 thirds, which is that number, are 3 times 2, which are 6.

6. 5 is 1 half of what number?

7. If 1 fourth of a lemon cost 2 cents, what will 1 cost?

8. If 1 fourth of a melon cost 5 cents, what will 1 cost?

9. 3 is 1 fourth of what number?

10. 7 is 1 third of what number?

11. 12 is 1 fifth of what number?

12. 7 is a fourth of what number?

13. What will 4 fifths of a pound of cinnamon cost, if 1 fifth of a pound cost 5 cents?

14. If 1 fifth of a yard of cloth cost $2, what will a yard cost?

15. If 1 sixth of a gallon of vinegar cost 2 cents, what will 1 gallon cost?

16. A man, being asked the value of his horse, said, 1 eighth of its value is $12; what was its value?

17. A man gave 15 cents for his lodging, which was 1 seventh as much as his breakfast cost; how much did his breakfast cost?

18. Bought 1 eighth of a yard of cloth for 4 dimes; what will a yard cost at this rate?

19. If 1 tenth of a yard of cloth cost 47 cents, how much is the price of a yard?

20. What will 1 yard of cloth cost, if 1 ninth of a yard cost 5 cents?

21. What will 1 bushel of corn cost, if 1 seventh of a bushel cost 5 cents?

22. What will a hogshead of molasses cost, if 1 eighth of a hogshead cost $3?

23. What will be the cost of 2 cords of wood, if 1 eleventh of a cord cost 30 cents?

24. If 1 twelfth of the distance from Albany to Wilbraham is 9 miles, what is the entire distance?

25. 9 is 1 tenth of what number?

26. 15 is 1 seventh of what number?

27. 16 is 1 fifth of what number?

28. 12 is 1 fifth of 6 times what number?

29. 15 is 1 sixth of 5 times what number?

30. 18 is 1 fourth of 6 times what number?

31. 10 is 1 eighth of 20 times what number?

32. 15 is 1 seventh of 5 times what number?

33. 20 is 1 eighth of 16 times what number?

34. 30 is 1 third of 6 times what number?

35. A boy's hat cost $3, which was 1 fifth of the cost of his coat; the cost of the coat is required.

36. Mr. B.'s saddle cost $9, which was 1 fortieth of 6 times the cost of his horse; the cost of the horse is required.

37. Henry gave 5 cents for a piece of pie, which was 1 twentieth of 4 times as much as his breakfast cost; what was the cost of his breakfast?

38. A man, being asked his age, answered that his youngest son's age, which was 12 years, was just 1 twelfth of 3 times his age; required the father's age.

39. Mrs. B.'s shawl cost $9, which was 1 tenth of 3 times the cost of her dress; what was the cost of her dress?

40. John said to James, who is now 10 years old, your age is 1 eighth of 4 times my age; how old is John?

LESSON XXIX.

1. If 2 thirds of a melon cost 4 cents, what will 1 melon cost?

ANALYSIS.—If 2 thirds of a melon cost 4 cents, 1 third will cost 1 half of 4 cents, which is 2 cents, and 3 thirds, which is 1 melon, will cost 3 times 2 cents, which are 6 cents.

2. If 2 thirds of an orange cost 5 cents, what will 1 orange cost?
3. If 3 fourths of a pound of cotton cost 9 cents, what will 1 pound cost?
4. If 4 thirds of a pound of spice cost 16 cents, what will 1 pound cost?
5. If 3 fourths of a pound of cinnamon cost 12 cents, what will 1 pound cost?
6. If \$4 will buy 2 fifths of a barrel of fish, what will 1 fifth of a barrel cost?
7. What will 1 yard of cloth cost, if 4 sixths of a yard cost 120 cents?
8. What will 1 hogshead of molasses cost, if 5 sevenths of a hogshead cost \$15?
9. 8 is 2 thirds of what number?

ANALYSIS.—If 2 thirds of some number is 8, 1 third of that number is 1 half of 8, which is 4; and 3 thirds, which is that number, are 3 times 4, which are 12. Therefore, 8 is 2 thirds of 12.

REMARK.—Representing the conditions and solutions of questions by symbols, will aid young pupils in comprehending the more difficult parts of arithmetical analysis. The condition and analysis of the preceding question may be represented, thus:

If 

$= \frac{2}{3}$ of some number $= 8$,

$= \frac{1}{3}$ of that number $= 4$,

and 

$= \frac{3}{3}$, which is that number, $=12$.

10. 12 is 2 thirds of what number?
11. 4 is 2 thirds of what number?
12. 10 is 2 sevenths of what number?
13. 9 is 3 fourths of what number? -
14. 12 is 3 fourths of what number?
15. 12 is 6 elevenths of what number?
16. 14 is 7 eighths of what number?
17. 14 is 2 sevenths of what number?
18. 6 is 3 tenths of what number?
19. 9 is 3 sevenths of what number?
20. 15 is 5 sixths of what number?
21. 15 is 3 halves of what number?
22. 18 is 9 eighths of what number?
23. 20 is 5 thirteenths of what number?
24. 24 is 8 fifths of what number?
25. 26 is 13 ninths of what number?
26. 2 thirds of 12 is 2 fifths of what number?
27. 3 fourths of 12 is 3 eighths of what number?
28. 3 fourths of 8 is 2 sevenths of what number?
29. 3 fifths of 25 is 5 fourths of what number?
30. 2 sevenths of 14 is 4 ninths of what number?
31. 4 sevenths of 21 is 3 tenths of what number?
32. 2 thirds of 15 is 5 fourths of what number?
33. 7 eighths of 48 is 3 halves of what number?
34. 8 ninths of 36 is 4 fifths of what number?
35. 7 thirds of 18 is 3 fifths of what number?
36. A watch cost $16, and 3 halves of the cost of the watch was 8 thirds of the cost of the chain; what was the cost of the chain?
37. A horse was sold for $96, which was 8 sevenths of what it cost; what was the cost of the horse?
38. In a certain school there are 36 girls, and 5 fourths of the number of girls equals 3 fifths of the number of boys; how many boys were there in the school?
39. Mary is 14 years old, and 4 sevenths of her age is 2 thirds of Hezekiah's age; what is his age?

40. A piece of cloth containing 12 yards was sold for $60, which was 5 fourths of the cost; how much did it cost, and what was the gain on each yard?

41. A. has 48 geese, and 3 fourths of his number is equal to 9 sevenths of B.'s number; how many geese has B.?

42. The head of a fish is 12 inches long, and 3 fourths of the length of the head is 3 fifteenths of the length of the body; required the length of the fish.

43. A farm was sold for $1200, which was only 6 sevenths of what it was worth; how much was lost by the bargain?

44. $48 is 3 fifths of the cost of 12 yards of cloth; for what must it be sold a yard to gain $16 on the whole?

45. A man gave $60 for a suit of clothes, which was 1 fifth of his yearly income; 1 sixth of the remainder he spent for a watch, and what then remained was 4 fifths of his brother's yearly income. What was the yearly income of each?

LESSON XXX.

1	half	is	written	thus,	$\frac{1}{2}$.
1	third	is	written	thus,	$\frac{1}{3}$.
1	fourth	is	written	thus,	$\frac{1}{4}$.
1	sixth	is	written	thus,	$\frac{1}{6}$.
1	seventh	is	written	thus,	$\frac{1}{7}$.
2	thirds	is	written	thus,	$\frac{2}{3}$.
3	fourths	is	written	thus,	$\frac{3}{4}$.
7	eighths	is	written	thus,	$\frac{7}{8}$.
9	tenths	is	written	thus,	$\frac{9}{10}$.
5	sevenths	is	written	thus,	$\frac{5}{7}$.
2	fifths	is	written	thus,	$\frac{2}{5}$.
5	thirds	is	written	thus,	$\frac{5}{3}$. &c.

REMARK.—The above expressions are called *fractions*. The figure above the short horizontal line is called the *numerator*, and the figure below the line is called the *denominator*. For example, in the fraction $\frac{3}{4}$, the 3 is the numerator, and the 4 is the denominator.

The *denominator* of a fraction shows into how many equal parts the thing is divided; and the *numerator* shows how many of these parts are taken.

1. If you cut an orange into 3 equal parts, what is 1 of these parts called?

2. If a lemon be cut into 4 equal pieces, what will 1 of these pieces be called? 2 pieces? 3? 4?

3. If a bushel of apples be divided into 6 equal parts, what will 1 of these parts be called? 3? 4? 6?

4. If a basket of peaches be divided in 8 equal parts, what will 3 of these parts be called? 5? 6? 7?

5. How can you find 2 thirds of an apple?

6. How can you find 3 fourths of an orange?

7. In $\frac{4}{2}$ how many times 1?

•ANALYSIS.—In 1 there are 2 halves; therefore, 1 half the number of halves equals the number of ones. 1 half of 4 is 2; therefore, $\frac{4}{2}$ equals 2.

8. In $\frac{6}{2}$ how many times 1?

9. In $\frac{8}{2}$ how many times 1?

10. In $\frac{10}{2}$ how many times 1?

11. In $\frac{12}{2}$ how many times 1?

12. In $\frac{14}{2}$ how many times 1?

13. In $\frac{6}{3}$ how many times 1?

14. In $\frac{12}{3}$ how many times 1?

15. In $\frac{9}{3}$ how many times 1?

16. In $\frac{15}{3}$ how many times 1?

17. In $\frac{18}{2}$ how many times 1?

18. In $\frac{18}{3}$ how many times 1?

19. In $\frac{21}{2}$ how many times 1?

20. In $\frac{8}{4}$ how many times 1?

21. In $\frac{12}{4}$ how many times 1?

22. In $\frac{16}{4}$ how many times 1 ?
23. In $\frac{20}{4}$ how many times 1 ?
24. In $\frac{24}{4}$ how many times 1 ?
25. In $\frac{36}{6}$ how many times 1 ?
26. In $\frac{6}{6}$ how many times 1 ?
27. In $\frac{10}{5}$ how many times 1 ?
28. In $\frac{25}{5}$ how many times 1 ?
29. In $\frac{15}{6}$ how many times 1 ?
30. In $\frac{55}{6}$ how many times 1 ?
31. In $\frac{60}{5}$ how many times 1 ?
32. In $\frac{18}{6}$ how many times 1 ?
33. In $\frac{30}{9}$ how many times 1 ?
34. In $\frac{48}{8}$ how many times 1 ?
35. In $\frac{50}{10}$ how many times 1 ?
36. In $\frac{45}{15}$ how many times 1 ?
37. In $\frac{72}{18}$ how many times 1 ?
38. In $\frac{100}{2}$ how many times 1 ?
39. In $\frac{50}{7}$ how many times 1 ?
40. In $\frac{63}{9}$ how many times 1 ?

REMARK.—This is called reducing fractions to *whole* or *mixed* *numbers*. A mixed number is a whole number with a fraction added to it. Thus, $3\frac{1}{2}$ is a *mixed number.*

When the *numerator* is less than the denominator, the value is less than a unit, and the expression is called a *proper fraction.* When the *numerator* is equal to, or greater than the denominator, the value is equal to, or greater than a unit, and the expression is called an *improper fraction.*

41. What kind of a fraction is it called, when the *numerator* is less than the denominator?

42. What kind of a fraction is it called, when the *denominator* is greater than the numerator?

43. When is the value of a fraction *greater* than a unit?

44. When the *denominator* is less than the numerator, what kind of a fraction is it called?

45. What kind of a fraction is it called, when the *numerator* is *larger* than the denominator?

46.   Reduce $\frac{7}{2}$ to   a   mixed   number.
47.   Reduce $\frac{9}{2}$ to   a   mixed   number.
48.   Reduce $\frac{15}{4}$ to   a   mixed   number.
49.   Reduce $\frac{19}{3}$ to   a   mixed   number.
50.   Reduce $\frac{23}{4}$ to   a   mixed   number.
51.   Reduce $\frac{21}{5}$ to   a   mixed   number.
52.   Reduce $\frac{57}{4}$ to   a   mixed   number.
53.   Reduce $\frac{89}{12}$ to   a   mixed   number.
54.   Reduce $\frac{94}{8}$ to   a   mixed   number.
55.   Reduce $\frac{25}{3}$ to   a   mixed   number.
56.   Reduce $\frac{37}{9}$ to   a   mixed   number.
57.   Reduce $\frac{47}{4}$ to   a   mixed   number.
58.   Reduce $\frac{78}{7}$ to   a   mixed   number.
59.   Reduce $\frac{37}{6}$ to   a   mixed   number.
60.   Reduce $\frac{34}{8}$ to   a   mixed   number.

LESSON XXXI.

1. James had $\frac{3}{4}$ of an apple, and John gave him $\frac{1}{4}$ more; how many had he then ?

2. Mary had $\frac{5}{7}$ of an orange, and her father gave her $\frac{2}{7}$ of an orange more; how many had she then ?

3. Robert had $\frac{2}{3}$ of a melon, and bought $\frac{2}{3}$ of another; how many had he then ?

4. Susan had $\frac{5}{7}$ of a pint of walnuts, and gave $\frac{3}{7}$ of a pint to her sister; how much had Susan left ?

5. James bought $\frac{19}{4}$ of a pound of candy, and on his way home ate $\frac{3}{4}$ of a pound; how much had he left ?

6. John gave $\frac{1}{5}$ of a pound of raisins to James, $\frac{4}{5}$ of a pound to Mary, and kept $\frac{3}{5}$ of a pound himself; how many fifths had he at first ?

7. Mortimer gave $\frac{4}{5}$ of a dollar for a hat, $1\frac{2}{5}$ for a vest, and had $3\frac{4}{5}$ remaining; how much had he at first ?

8. Jane had 5 pounds of cinnamon, and Harriet had $2\frac{3}{4}$ pounds; how many more had Jane than Harriet ?

9. What is the sum of $\frac{2}{6}$ of a dollar, $\frac{3}{6}$ of a dollar, and $\frac{1}{6}$ of a dollar?

10. $\frac{8}{9} + \frac{4}{9}$ are how many?

11. $\frac{7}{8} + \frac{6}{8}$ are how many?

12. $\frac{5}{8} + \frac{6}{8}$ are how many?

13. $\frac{9}{10} + \frac{8}{10}$ are how many?

14. $\frac{3}{4} + \frac{2}{4}$ are how many?

15. $1\frac{6}{3} + \frac{4}{3}$ are how many?

16. $1\frac{4}{3} + \frac{5}{3}$ are how many?

17. $1\frac{4}{8} + 1\frac{6}{8}$ are how many?

18. $\frac{9}{4} + 1\frac{3}{4}$ are how many?

19. $\frac{7}{8} + \frac{6}{8} + \frac{5}{8} =$ how many?

20. $\frac{3}{8} + \frac{6}{8} + \frac{4}{8} =$ how many?

21. $\frac{7}{10} + \frac{8}{10} + \frac{9}{10} =$ how many?

22. $\frac{2}{7} + \frac{4}{7} + \frac{6}{7} =$ how many?

23. $\frac{8}{9} - \frac{4}{9}$ are how many ninths?

24. $2\frac{7}{8} - 1\frac{7}{8}$ are how many?

25. $1\frac{9}{4} - \frac{3}{4}$ are how many?

26. $2\frac{3}{3} - \frac{2}{3}$ are how many?

27. $3\frac{7}{5} - \frac{3}{5}$ are how many?

28. $1\frac{4}{3} - \frac{2}{3}$ are how many?

29. $3\frac{7}{8} - \frac{5}{8}$ are how many?

30. $1\frac{9}{7} - \frac{2}{7}$ are how many?

31. $4\frac{6}{9} - 1\frac{0}{9}$ are how many?

32. $4\frac{7}{8} - \frac{6}{8}$ are how many?

33. $5\frac{7}{12} - 3\frac{7}{12}$ are how many?

34. $\frac{9}{4} + \frac{7}{4} - \frac{3}{4} =$ how many?

35. $\frac{7}{8} + \frac{5}{8} - \frac{2}{8} =$ how many?

36. $\frac{4}{5} + 1\frac{7}{5} - \frac{9}{5} =$ how many?

37. $1\frac{4}{15} - 1\frac{0}{15} + 1\frac{2}{15} =$ how many?

38. $\frac{3}{5}$ of $60 - \frac{3}{4}$ of $24 =$ how many?

39. $\frac{7}{8}$ of $40 - \frac{2}{5}$ of $10 =$ how many?

40. $\frac{4}{5}$ of $15 + \frac{2}{3}$ of $9 - \frac{3}{4}$ of $12 =$ how many?

LESSON XXXII.

1. At $\frac{2}{3}$ of a cent each, what will 2 apples cost?

   ANALYSIS.—If 1 apple cost $\frac{2}{3}$ of a cent, 2 apples will cost twice $\frac{2}{3}$ of a cent, which are $\frac{4}{3}$, or $1\frac{1}{3}$ cents.

2. At $\frac{3}{8}$ of a cent each, what will 5 apples cost?
3. At $\frac{4}{5}$ of a dime a pound, what will 10 pounds of sugar cost?
4. At $1\frac{3}{4}$ dimes a pound, how many cents will 8 pounds of starch cost?
5. At $\frac{2}{5}$ of a cent each, what will 25 filberts cost?
6. At $\frac{8}{9}$ of a dime each, how many cents will 8 chickens cost?
7. At $\frac{2}{5}$ of a dollar a yard, what will 15 yards of linen cost?
8. If a man spend $\frac{3}{4}$ of a dollar a day, how much, at this rate, will he spend in 23 days?
9. If a man receive $\frac{2}{3}$ of an Eagle in a week, how many dollars will he receive in 52 weeks?
10. If 1 pound of sugar cost $1\frac{1}{3}$ dimes, what will 12 pounds cost?
11. At $5\frac{2}{3}$ cents a pound, what will 6 pounds of soap cost?
12. At $9\frac{3}{4}$ cents a pound, what will 8 pounds of pork cost?
13. At $6\frac{2}{5}$ cents each, what will 12 lemons cost?
14. At $7\frac{3}{5}$ cents each, what will 20 rabbits cost?
15. At $12\frac{1}{2}$ cents a dozen, what will 4 doz. eggs cost?
16. At $11\frac{2}{3}$ cents a pound, what will 6 pounds of honey cost?
17. At $\$7\frac{1}{4}$ a bbl., what will 10 bbl. of tobacco cost?
18. At $\$9\frac{3}{4}$ a bbl., what will 10 bbl. of pork cost?
19. What will 6 boxes of raisins cost, at $\$3\frac{2}{3}$ a box?

20. What will 14 bushels of wheat cost, at \$$1\frac{2}{3}$ a bushel?
21. What will 7 barrels of cider cost, at \$$3\frac{3}{4}$ a barrel?
22. If a barrel of flour cost \$4, what will $5\frac{3}{4}$ bbl. cost?
23. 5 times 4 and $\frac{3}{4}$ of 4 are how many?
24. 7 times 6 and $\frac{4}{6}$ of 6 are how many?
25. 9 times 7 and $\frac{5}{7}$ of 7 are how many?
26. 12 times 9 and $\frac{8}{9}$ of 9 are how many?
27. 5 times 10 and $\frac{4}{5}$ of 10 are how many?
28. 13 times 4 and $\frac{3}{4}$ of 4 are how many?
29. 8 times 7 and $\frac{3}{7}$ of 7 are how many?
30. 10 times 13 and $\frac{11}{13}$ of 13 are how many?
31. 7 times 20 and $\frac{3}{5}$ of 20 are how many?
32. How many are 4 times $\frac{2}{3}$?
33. How many are 4 times $2\frac{2}{3}$?
34. How many are 3 times $4\frac{3}{4}$?
35. How many are 5 times $3\frac{2}{7}$, $+\frac{5}{7}$?
36. How many are 7 times $9\frac{2}{3}$, $+\frac{2}{3}$?
37. How many are 8 times $12\frac{5}{7}$, $-\frac{5}{7}$?
38. How many are 9 times $10\frac{3}{4}$, $-\frac{3}{4}$?
39. How many are 6 times $12\frac{2}{3}$, $+2\frac{1}{3}$?
40. How many are 12 times $9\frac{7}{8}$, $+\frac{4}{8}$?

LESSON XXXIII.

1. If you give to 6 persons each $\frac{2}{3}$ of a dollar, how many dollars will be given to all?
2. What will be the cost of 4 yards of cloth, at $\frac{3}{4}$ of a dollar a yard?
3. If 1 yard of cloth cost \$$1\frac{2}{5}$, what will 10 yards cost?
4. How many oranges will it require to give to each of 9 boys $1\frac{1}{3}$ oranges?

5. How many barrels of flour does that man give away, who gives to each of 12 beggars $\frac{2}{3}$ of a barrel?

6. Anthony gave to each of his 7 companions $\frac{2}{3}$ of a pound of candy, and had $\frac{1}{3}$ of a pound left; how many pounds had he at first?

7. Thornton gave to each of 9 beggars $\frac{5}{6}$ of a dollar, and had $7 remaining; how much had he at first?

8. James gave $\frac{1}{12}$ of an orange to Jackson, $\frac{5}{12}$ to Joseph, and $\frac{3}{12}$ to John; what part of an orange had James remaining?

9. Harmon meeting 3 poor women and 5 poor men, gave to each woman $\frac{2}{3}$ of a dollar, and to each man $\frac{4}{5}$ of a dollar, and had only $4 remaining; how much had he at first?

10. How many quarts of chestnuts must that boy have, who gives to each of 20 persons $\frac{1}{5}$ of a quart, and has 7 quarts remaining?

11. Mary, after giving to each of her 12 companions as many pinks as she had roses, which were 2, had no flowers remaining but her roses; how many flowers had Mary at first?

12. What will 1 quart of vinegar cost, if 1 pint cost $\frac{3}{4}$ of a cent?

13. If 1 gill of molasses cost $\frac{2}{3}$ of a cent, what will 2 quarts cost?

14. If 2 pints of beans cost 4 cents, what will 1 peck cost?

15. If 3 pecks of buckwheat cost 96 cents, what will 1 pint cost?

16. What will 10 yards of silesia cost, if 1 yard cost $18\frac{3}{4}$ cents?

17. How many cents will $4\frac{2}{3}$ yards of silk cost, if 1 yard cost 6 dimes?

18. What will $\frac{3}{4}$ of a yard of muslin cost, if 1 yard cost 10 cents?

19. What will 7 spools of thread cost, if 1 spool cost $7\frac{5}{8}$ cents?

20. What will $8\frac{2}{3}$ yards of silk cord cost, at 6 cents a yard?

21. If a boat sail 1 mile in 5 minutes, in what time, at the same rate, will it sail $9\frac{1}{4}$ miles?

22. What will $6\frac{3}{4}$ yards of muslin cost, if 1 yard cost 8 cents?

23. What will 5 pictures cost, at $8\frac{3}{8}$ cents each?

24. How much will $9\frac{3}{4}$ barrels of cider cost, at $4 a barrel?

25. What will 12 hats cost, at $3\frac{3}{4}$ each?

26. What will $5\frac{3}{4}$ melons cost, at 4 dimes each?

27. What will be the cost of 13 yards of bishop lawn at $1\frac{4}{13}$ a yard?

28. What will be the cost of 8 glasses, at $15\frac{3}{4}$ each?

29. What amount of money will be required to purchase 30 pounds of rice, at $6\frac{3}{5}$ cents a pound?

30. What will be the cost of 23 pounds of crackers, at $8\frac{1}{2}$ cents a pound?

31. What will 9 barrels of fish cost, at $12\frac{2}{3}$ a barrel?

32. If a man's income is $9\frac{1}{2}$ dimes in 1 hour, how much will it be in 24 hours?

33. If 1 gold pen cost $2\frac{2}{3}$, how much will 6 cost?

34. How many pounds of meat, at 5 cents a pound, can you buy for $3\frac{2}{3}$?

35. What will be the cost of 3 quarts of nuts, at 64 cents a peck?

36. If railroad fare is $5\frac{1}{2}$ cents a mile, what is the price of a ticket for 12 miles?

37. How many dollars, dimes, and cents will 12 yards of cloth cost, at 62 cents a yard?

38. How many dollars and cents will 4 pecks of grass-seed cost, if 1 pint cost 5 cents?

39. How much is 13 pounds of cotton worth, at $13\frac{3}{4}$ cents a pound?

40. How many cents will 16 bushels of potatoes cost, at $2\frac{5}{8}$ dimes a bushel?

LESSON XXXIV.

An *integer* is any whole number.

There are two methods of multiplying a fraction by an integer:

RULE I.—Multiply the numerator by the integer.

RULE II.—Divide the denominator by the integer.

NOTE.—See Stoddard's New Practical Arithmetic, page 95, Arts. 140 and 141.

1. How many are 5 times $\frac{7}{10}$?

    ANALYSIS 1ST.—5 times $\frac{7}{10}$ are $\frac{35}{10}$, or $3\frac{1}{2}$.

    ANALYSIS 2D.—5 times $\frac{7}{10}$ are $\frac{7}{2}$, or $3\frac{1}{2}$.

2. How many are 3 times $\frac{13}{9}$?
3. How many are 9 times $\frac{6}{27}$?
4. How many are 5 times $\frac{27}{15}$?
5. How many are 6 times $\frac{27}{12}$?
6. How many are 9 times $\frac{4}{18}$?
7. How many are 9 times $\frac{18}{36}$?
8. How many are 7 times $\frac{39}{21}$?
9. 8 times $\frac{13}{16}$ are how many?
10. 11 times $\frac{18}{22}$ are how many?
11. 13 times $\frac{15}{26}$ are how many?
12. 2 times $\frac{9}{4}$ are how many?
13. 5 times $\frac{17}{20}$ are how many?
14. 6 times $\frac{40}{12}$ are how many?
15. 7 times $\frac{34}{14}$ are how many?
16. 12 times $\frac{18}{36}$ are how many?
17. How many times 5 are 8 times $\frac{40}{16}$?
18. How many times 12 are 9 times $\frac{48}{18}$?
19. How many times 8 are 11 times $\frac{64}{22}$?
20. How many times 100 are 25 times $\frac{400}{50}$?
21. How many times 20 are 35 times $\frac{800}{70}$?
22. 5 times $\frac{48}{6}$ is 4 times Mary's age; what is her age?

23. 13 times $\frac{150}{39}$ equals $\frac{1}{2}$ of the number of dollars a wagon cost; required the cost of the wagon.

24. 25 times $\frac{100}{26}$ equals $\frac{1}{200}$ of the number of men the Mexican General Santa Anna had at the battle of Buena Vista; how many men had he?

25. 6 times $\frac{200}{12}$ is $\frac{1}{15}$ of the number of men he had wounded; how many men were wounded?

26. 7 times $\frac{560}{14}$ is $\frac{1}{4}$ of the number of men he had killed; how many men were killed?

27. 4 times $\frac{450}{8}$ is $\frac{1}{20}$ of the number of men General Taylor had at the same battle; how many had he?

28. 9 times $\frac{67}{9}$ is $\frac{1}{4}$ of the number of men he had killed; how many were killed?

29. 8 times $\frac{15}{16}$ is $\frac{1}{2}$ of how many times 3?

30. 4 times $\frac{25}{12}$ is $\frac{1}{3}$ of how many times 5?

31. A boy worked 12 months, at the rate of $10\frac{1}{4}$ a month; how much did his year's wages amount to?

32. If 2 quarts of beans cost 48 cents, what will 1 gill cost?

33. What will 3 oranges cost, at $\frac{3}{4}$ of a cent each?

34. A piece of work was performed in 96 hours, by working 6 hours a day; how many days did it take?

35. If 1 man can dig a ditch in 15 days, how long will it take 5 men to dig it?

36. If a certain quantity of provision serve a family of 4 persons 16 days, how long would it last a family of 8 persons?

37. If 8 men can perform a piece of work in 56 days, in how many days can 112 men do the same?

38. If 3 men can plow 18 acres in 6 days, in how many days could 9 men do the same?

39. 4 men can mow a certain field in $6\frac{1}{4}$ days, in how many days could 5 men perform the same work?

40. How many kegs of butter, at $4 a keg, will pay for 6 barrels of cider, at $3\frac{3}{8}$ a barrel?

41. A merchant bought 6 yards of cloth, and sold it for $20, which was $\frac{10}{9}$ of what it cost; what did it cost a yard?

42. Bought 36 yards of cloth, and sold $\frac{5}{6}$ of it for $25, which was $\frac{5}{4}$ of what it cost; how much would I have gained by selling the whole at the same rate?

43. 7 men in $\frac{5}{6}$ of a day can earn $10; how long would it take 1 man to earn the same?

44. James is $3\frac{2}{3}$ years of age, which is $\frac{1}{6}$ of the age of Henry; and Henry is 9 times as old as George; what is the age of Henry and of George respectively?

45. $\frac{1}{3}$ of 36 is 3 times $\frac{1}{2}$ of what number?

46. $\frac{1}{2}$ of 32 is $\frac{2}{3}$ of 3 times what number?

47. $\frac{2}{3}$ of 60 is $\frac{2}{4}$ of 2 times what number?

48. $\frac{1}{4}$ of 40 is $\frac{3}{4}$ of as many dollars as Mr. B.'s horse cost; what was the cost of his horse?

49. A person, being asked his age, said, that $\frac{3}{4}$ of 80 was $\frac{2}{3}$ of 10 times his age; what was his age?

50. Morgan is 20 years old, and $\frac{4}{5}$ of his age is $\frac{4}{5}$ of the age of his brother; what is his brother's age?

LESSON XXXV.

1. How many thirds are there in 3?

> ANALYSIS 1st.—In 1 there are 3 thirds, and in 3 there are 3 times 3 thirds, which are $\frac{9}{3}$.
>
> ANALYSIS 2D.—In 1 there are 3 thirds; therefore, 3 times the number of whole ones equal the number of thirds. 3 times 3 are 9. Therefore, in 3 there are $\frac{9}{3}$.

2. How many fourths are there in 3?

3. How many halves are there in 6?

4. How many fifths are there in 4? in 5?

5. How many fifths are there in 7? in 8?

6. How many sixths are there in 4 ? in 3 ? in 5 ?

7. How many sevenths are there in 2 ? in 4 ? in 6 ?

8. How many eighths are there in 7 ? in 4 ? in 5 ?

9. How many fifteenths are there in 2 ? in 3 ? in 6 ?

10. How many tenths are there in 4 ? in 6 ? in 7 ?

11. How many fourths are there in 3 and $\frac{3}{4}$?

12. How many thirds are there in 4 and $\frac{1}{3}$?

13. How many thirds are there in 3 and $\frac{2}{3}$?

14. How many halves are there in 8 and $\frac{1}{2}$?

15. Reduce $6\frac{5}{8}$ to an improper fraction.

16. Reduce $9\frac{3}{6}$ to an improper fraction.

17. Reduce $7\frac{3}{7}$ to an improper fraction.

18. Reduce $5\frac{3}{4}$ to an improper fraction.

19. Reduce $4\frac{3}{8}$ to an improper fraction.

20. Among how many men can $5\frac{4}{7}$ bales of cotton be distributed, provided each receives $\frac{1}{7}$ of a bale ?

21. Among how many boys can $7\frac{3}{8}$ oranges be divided, provided each receives $\frac{1}{8}$ of an orange ?

22. 8 and $\frac{4}{6}$ are how many times $\frac{2}{9}$?

ANALYSIS.—8 and $\frac{4}{6}$ equal $\frac{74}{9}$. $\frac{2}{9}$ is contained in $\frac{74}{9}$ 37 times.

23. $9\frac{6}{8}$ are how many times $\frac{2}{8}$?

24. $9\frac{2}{6}$ are how many times $\frac{3}{6}$?

25. $7\frac{1}{6}$ are how many times $\frac{3}{6}$?

26. $12\frac{2}{9}$ are how many times $\frac{5}{9}$?

27. $7\frac{5}{7}$ are how many times $\frac{3}{7}$?

28. $7\frac{3}{4}$ are how many times $\frac{3}{4}$?

29. $4\frac{3}{6}$ are how many times $\frac{9}{6}$?

30. $10\frac{4}{5}$ are how many times $\frac{3}{5}$?

31. $8\frac{6}{9}$ are how many times $\frac{3}{9}$?

32. $12\frac{5}{7}$ are how many times $\frac{3}{7}$?

33. 4 times $3\frac{1}{3}$ are how many times $\frac{2}{3}$?

34. 9 times $1\frac{2}{6}$ are how many times $\frac{4}{6}$?

35. 5 times $6\frac{2}{5}$ are how many times $\frac{2}{5}$?

36. 8 times $8\frac{2}{3}$ are how many times $\frac{2}{3}$?
37. 6 times $2\frac{4}{7}$ are how many times $\frac{3}{7}$?
38. A boy distributed $9\frac{3}{5}$ apples equally among his companions, giving to each $\frac{4}{5}$ of an apple; how many companions had he?
39. Homer distributed \$$12\frac{4}{5}$ equally among some poor women, giving to each \$$1\frac{3}{5}$; how many women were there?
40. Mary gave $\frac{3}{5}$ of a pie to each of her 9 visitors; how many pies did it take?
41. Bought 8 yards of cloth, at \$$5\frac{2}{3}$ a yard; how much silk, worth \$$1\frac{1}{3}$ a yard, will pay for it?
42. Bought 9 yards of cloth, worth \$$1\frac{5}{6}$ a yard, and paid for it with raisins, at \$$1\frac{1}{2}$ a box; how many boxes did it take?
43. How many bushels of turnips, at \$$\frac{2}{5}$ a bushel, can be bought for 8 bushels of apples, at \$$\frac{3}{5}$ a bushel?
44. How many apples, at $\frac{3}{4}$ of a cent each, can be bought for 6 oranges, at $2\frac{1}{4}$ cents each?
45. How many yards of cloth, at \$$\frac{5}{7}$ a yard, can be bought for 10 firkins of butter, at \$$5\frac{3}{7}$ a firkin?
46. How many geese, at \$$\frac{7}{8}$ each, can be bought for 14 ducks, at \$$\frac{3}{8}$ each?
47. How many boxes of cheese, worth \$$2\frac{3}{7}$ a box, may be had for 17 firkins of butter, at \$$1\frac{3}{7}$ a firkin?
48. How many barrels of flour, at \$$5\frac{2}{3}$ a barrel, may be had for 17 bundles of cotton, at \$$1\frac{1}{3}$ a bundle?
49. How many sheep, at \$$1\frac{1}{3}$ a head, may be had for 8 calves, at \$$3\frac{2}{3}$ each?
50. How many quarts of pease, at $\frac{2}{3}$ of a cent a pint, may be had for 12 qt. of molasses, at $4\frac{1}{3}$ cents a qt.?
51. Mary and Jane together picked 5 times $2\frac{2}{15}$ quarts of strawberries, and shared them equally with their companions; how many companions had they, provided each received $1\frac{1}{3}$ quarts?

LESSON XXXVI.

LESSONS XXXV AND XXXVI COMBINED.

There are two methods of dividing a fraction by an integer:
RULE 1ST.—Divide the numerator by the integer.
RULE 2D.—Multiply the denominator by the integer.

☞ See Stoddard's New Practical Arithmetic, pages 95, 96.

1. If 2 yards of cloth cost $$\frac{4}{5}$$, what will 1 yard cost?

ANALYSIS.—If 2 yards of cloth cost $\frac{4}{5}$ of a dollar, 1 yard will cost $\frac{1}{2}$ of $\frac{4}{5}$ of a dollar, which is $\frac{2}{5}$ of a dollar.

2. If 3 apples cost $\frac{6}{7}$ of a cent, what will 1 apple cost?

3. If 2 oranges cost $\frac{6}{7}$ of a cent, what will 1 orange cost?

4. If 3 yards of cord cost $\frac{12}{13}$ of a cent, what will 1 yard cost?

5. If 2 pounds of sugar cost $8\frac{2}{3}$ (or $\frac{26}{3}$ cents), what will 1 pound cost?

6. If 2 pine-apples cost $14\frac{2}{3}$ cents, what will 1 pine-apple cost?

7. If $\frac{2}{3}$ of a melon is worth 2 oranges, how much is 1 orange worth?

8. If 3 apples are worth $2\frac{2}{7}$ quinces, what is 1 apple worth?

9. How many times 7 are $\frac{21}{23}$?

10. How many times 13 are $7\frac{4}{5}$?

11. How many times 7 are $9\frac{1}{3}$?

12. How many times 21 are $16\frac{4}{1}$?

13. How many times 8 are $33\frac{3}{5}$?

14. How many times 7 are $10\frac{1}{2}$?

15. How many times 11 are $40\frac{1}{3}$?

16. How many times 18 are $14\frac{2}{5}$?

17. How many times 3 are $44\frac{4}{5}$?

18. How many times 9 are $7\frac{5}{7}$?

19. How many times 6 are $6\frac{6}{8}$?

20. How many times 15 are $33\frac{1}{5}$?

21. If 9 oranges are worth $\$\frac{9}{10}$, how many cents is 1 orange worth?

22. If 4 boxes of figs cost $\frac{4}{5}$ of an Eagle, how many dollars will 1 box cost?

23. If 7 pounds of cheese cost $\$\frac{7}{10}$, how many cents will 1 pound cost?

24. If 3 cakes cost $\frac{9}{10}$ of a dime, how many cents will 1 cake cost?

25. If 4 pounds of chocolate cost $4\frac{1}{5}$ dimes, how many cents is that a pound?

26. What will 1 portfolio cost, if 3 cost $\frac{9}{10}$ of an Eagle?

27. If 8 quarts of alcohol cost 32 dimes, how many cents will 2 gills cost?

28. What will 1 pound of sugar cost, if 4 pounds cost $18\frac{2}{3}$ cents?

29. If 6 pounds of cheese cost $31\frac{1}{2}$ cents, what will 1 pound cost?

30. If 12 eggs cost $9\frac{2}{5}$ cents, what will be the cost of 1 egg?

31. If 7 yards of cloth cost $\$24\frac{1}{2}$, what will 1 yard cost?

32. If 5 silk shawls cost $\$27\frac{1}{2}$, how much will 1 cost?

33. If 9 pairs of boots cost $\$32\frac{2}{5}$, how much is that a pair?

34. If 9 oranges are worth $30\frac{2}{5}$ walnuts, how many walnuts is 1 orange worth?

35. A boy gave 8 apples for $18\frac{2}{3}$ marbles, how many marbles did he get for 1 apple?

36. A boy gave 7 cents for $17\frac{1}{2}$ crackers; how many did he get for 1 cent?

37. Mary gave 10 pins for $23\frac{1}{3}$ chestnuts; how many did she get for 1 pin?

38. If 3 yards of broadcloth are worth $18\frac{6}{7}$ yards of muslin, how many yards of muslin may be had for 1 yard of broadcloth?

39. If John can walk 13 miles while Josiah is walking 32½ miles, how far can Josiah walk while John is walking 1 mile?

40. If 2 chestnuts are worth $\frac{2}{10}$ of a cent, and 20 chestnuts are worth $\frac{4}{7}$ of a lemon, how many cents is 1 lemon worth?

41. If 2 oranges cost $\frac{4}{8}$ of a cent, what will 1 orange cost?

ANALYSIS.—If 2 oranges cost $\frac{4}{8}$ of a cent, 1 orange will cost $\frac{1}{2}$ of $\frac{4}{8}$ of a cent, which is $\frac{4}{8}$ of a cent.

42. If 3 yards of linen cost 1\frac{3}{8}$, what will 1 yard cost?

43. If 7 yards of tape cost 13$\frac{2}{8}$ cents, what will 1 yard cost?

44. If 2 pints of molasses cost 1$\frac{2}{6}$ dimes, how many cents will 1 gallon cost?

45. How many times 8 are 6$\frac{1}{2}$?

46. How many times 6 are 5$\frac{2}{3}$?

47. How many times 4 are 4$\frac{1}{2}$?

48. How many times 2 are 13$\frac{1}{3}$?

49. How many times 7 are 7$\frac{2}{5}$?

50. How many times 8 are 9 times $\frac{27}{18}$?

51. How many times 3 are 6 times 1$\frac{11}{18}$?

ANALYSIS.—1$\frac{11}{18}$ equals $\frac{29}{18}$. 6 times $\frac{29}{18}$ are $\frac{29}{3}$. 3 is contained in $\frac{29}{3}$, $\frac{29}{9}$, or 3$\frac{2}{9}$ times.

52. How many times 9 are 10 times 2$\frac{3}{20}$?

53. How many times 7 are 5 times 3$\frac{1}{10}$?

54. How many times 5 are 15 times 3$\frac{7}{30}$?

55. How many times 7 are 15 times 1$\frac{28}{45}$?

56. How many times 5 are 3 times 2$\frac{2}{9}$?

57. How many times 6 are 4 times 5$\frac{2}{8}$?

58. If 1$\frac{3}{8}$ yards of cloth are worth 2\frac{1}{2}$, what is 1 yard worth?

59. If 6$\frac{2}{8}$ bunches of grapes are worth 40 cents, how many cents is 1 bunch worth?

60. If 3$\frac{2}{8}$ baskets of peaches are worth 5\frac{1}{2}$, what is 1 basket of peaches worth?

LESSON XXXVII.

1. What is $\frac{1}{3}$ of 2?

ANALYSIS.—$\frac{1}{3}$ of 1 is $\frac{1}{3}$; if $\frac{1}{3}$ of 1 is $\frac{1}{3}$, $\frac{1}{3}$ of 2 is twice $\frac{1}{3}$, which are $\frac{2}{3}$. Therefore, $\frac{1}{3}$ of 2 is $\frac{2}{3}$ of 1.

2. What is $\frac{1}{3}$ of 4? of 8?
3. What is $\frac{1}{4}$ of 2? 3? 5? 7? 9?
4. What is $\frac{1}{5}$ of 3? 5? 7? 9? 11?
5. What is $\frac{1}{5}$ of 2? 3? 4? 7? 8?
6. What is $\frac{1}{6}$ of 2? 3? 5? 7? 9?
7. What is $\frac{1}{7}$ of 2? 3? 5? 4? 6? 9? 11?
8. What is $\frac{1}{8}$ of 2? 4? 3? 5? 6? 7? 9?
9. What is $\frac{1}{9}$ of 2? 4? 7? 6? 3? 12? 11?
10. What is $\frac{1}{10}$ of 7? 2? 4? 6? 9? 14? 15?
11. If 2 apples cost 3 cents, what will 1 apple cost?

ANALYSIS.—If 2 apples cost 3 cents, 1 apple will cost $\frac{1}{2}$ of 3 cents, which is $\frac{3}{2}$, or $1\frac{1}{2}$ cents.

12. If 2 apples cost 5 cents, what will 1 apple cost?
13. If 3 pens cost 8 cents, what will 1 pen cost?
14. If 3 yards of tape cost 14 cents, what will 1 yard cost?
15. If 5 barrels of flour cost $21, what will 1 barrel cost?
16. If 7 pecks of dried apples cost 23 dimes, what will 1 peck cost?
17. If 4 chickens cost 9 dimes, what will 1 chicken cost?
18. What will 1 pound of tamarinds cost, if 6 pounds cost 27 dimes?
19. What will 1 barrel of flour cost, if 3 barrels cost $25?
20. If you divide 7 bushels of wheat equally among 5 persons, how much will each receive?
21. Joshua had 13 marbles, and Lewis had $\frac{1}{2}$ as many + $\frac{1}{2}$ of a marble; how many had Lewis?

22. A. divided 3 barrels of flour equally among 11 families; what part of a barrel did each receive?

23. A farmer divided 5 bushels of rye equally among 7 of his poor neighbors; what part of a bushel did he give to each?

24. Calvin had 4 pints of nuts, and shared them equally with 6 of his companions; how much did each receive?

25. Margaret, having 7 quarts of raspberries, shared them equally with 8 of her playmates; what part of a quart did each receive?

26. What will 1 pound of prunes cost, if 5 pounds cost 48 dimes?

27. What cost 2 boxes of figs, at 20 dimes for 7 boxes?

28. What will 4 quarts of strawberries cost, if 9 quarts cost 7 dimes?

29. What is $\frac{1}{3}$ of 2?

30. If $\frac{1}{3}$ of 2 is $\frac{2}{3}$, what is $\frac{2}{3}$ of 2?

31. What is $\frac{2}{3}$ of 5?

ANALYSIS.—$\frac{1}{3}$ of 5 is $\frac{5}{3}$, and $\frac{2}{3}$ are 2 times $\frac{5}{3}$, which are $\frac{10}{3}$, or $3\frac{1}{3}$.

32. What is $\frac{2}{3}$ of 3? of 5? 6? 7? 9? 11?

33. What is $\frac{3}{5}$ of 2? of 3? 4? 9? 11? 13?

34. What is $\frac{5}{6}$ of 2? of 3? 5? 7? 9? 15?

35. What is $\frac{4}{7}$ of 3? of 5? 8? 9? 12? 20?

36. What is $\frac{7}{8}$ of 4? of 6? 8? 12? 11? 17?

37. What is $\frac{3}{8}$ of 3? of 5? 6? 9? 10? 15?

38. What is $\frac{3}{11}$ of 2? of 4? 5? 6? 10? 12?

39. What is $\frac{5}{12}$ of 2? of 4? 3? 9? 15? 21?

40. What is $\frac{7}{10}$ of 7? of 8? 9? 12? 15? 25?

41. How many cents will $\frac{2}{3}$ of a pound of candy cost, if 1 pound cost 2 dimes?

42. What will be the cost of $\frac{3}{4}$ of a box of raisins, if 1 box cost $3?

43. What will be the cost of $\frac{2}{3}$ of a yard of cloth, at 7 dimes a yard?

44. If a ton of hay cost $10, what will $2\frac{2}{3}$ tons cost?

45. Jeremiah is 91 years old, and $\frac{3}{7}$ of his age equals the age of his oldest son; how old is he?

46. Bought 24 yards of cloth for $48; but, being damaged, I sold $\frac{3}{4}$ of it, at $1\frac{5}{8}$ a yard, and the remainder for what it cost; how much did I lose?

47. Bought 14 yards of cloth, and sold $\frac{2}{7}$ of it, at $2 a yard, which amounted to $2 less than the whole cost; what did it cost a yard?

48. A horse was sold for $97, which was $1\frac{2}{7}$ times as much as it cost; what did the horse cost?

49. If 9 yards of cloth cost $17, what will 3 yards cost?

50. If 7 yards of cloth cost $25, what will 9 yards cost?

51. What will 2 pounds of opium cost, if 5 pounds cost $42?

52. If 5 pounds of indigo cost $32, what will 2 pounds cost?

53. A wagon was sold for $90, which was $\frac{3}{4}$ of what it cost; how much did it cost?

54. Two men started from the same place, and traveled the same way; one at the rate of 92 miles in 10 hours; the other at the rate of $1\frac{1}{10}$ miles in $\frac{1}{4}$ of an hour; how far apart will they be in 2 hours?

55. By a pipe $4\frac{5}{7}$ gallons of water ran into a cistern in 1 minute; how much did the vessel hold, provided it was filled in 9 minutes?

56. If 7 men can perform a piece of work in $13\frac{5}{7}$ days, how long would it take 4 men to do the same?

57. If 5 persons consume a barrel of flour in 9 weeks, what part of a barrel would they consume in 5 weeks?

58. If a man earn $\frac{7}{8}$ in a day, and a boy $\frac{5}{8}$, how much will both earn in 6 days?

59. Anthony spent $\frac{3}{4}$ of all his money, and the remainder he gave for 8 yards of cloth, at $2\frac{1}{4}$ a yard; how much had he at first?

60. From a piece of cloth a tailor cut 5 garments, each containing $3\frac{4}{7}$ yards; and there remained $2\frac{6}{7}$ yards; how many yards did the piece at first contain?

61. What will 9 pounds of rice cost, if 7 pounds cost 43 cents?

62. An individual, after spending $\frac{13}{17}$ of all his money, had only $40 remaining; how much had he at first?

63. An old lady bought 30 eggs, at the rate of 2 for 5 cents; what did they cost?

64. How much will 13 pounds of coffee cost, if 26 pounds cost $7?

65. What will 7 gallons of molasses cost, if 6 pints cost 27 cents?

66. If 5 lamps cost $7\frac{1}{2}$, what will 7 lamps cost?

67. If 5 horses can, in $4\frac{2}{5}$ days, consume 20 bushels of oats, in how many days can 11 horses consume the same?

68. If 15 gold pens cost $20, what will 5 of them cost?

69. If $\frac{1}{8}$ of an acre of land be worth $14, what are 10 acres worth?

70. $25 is $\frac{5}{7}$ of the cost of B.'s watch; what was the cost of his watch?

71. Mortimer's hat cost $5, and $\frac{2}{3}$ of the cost of his hat is $\frac{1}{12}$ of the cost of his coat; what was the cost of his coat?

72. If a man in $\frac{4}{15}$ of a day walk 8 miles, how far can he walk in 5 days?

73. From a piece of cloth containing 25 yards, a tailor cut 8 suits of clothes, each containing $2\frac{3}{4}$ yards; how many yards remained?

74. If a man can cut 1 cord of wood in 5 hours, how

many cords can he cut in 4 days, by working 12 hours a day ?

75. A man bought 7 sheep, at the rate of 9 for $\$5\frac{1}{2}$; what did they cost him ?

76. A boy bought 13 oranges, giving 9 apples for 3 oranges; how many apples did his oranges cost ?

77. If 25 cents buy 7 lemons, how many cents will 9 lemons cost ?

78. $\frac{3}{5}$ of 45 equals $\frac{3}{4}$ as many dollars as Andrew has; how many dollars has Andrew ?

79. $\$30\frac{6}{7}$ is $\frac{1}{7}$ of all the money A. had; how much money had A. ?

80. What will 3 pecks of flax-seed cost, if 3 pints cost 3 dimes ?

81. What will 1 quart of clover-seed cost, if 2 pecks cost $3 and 2 dimes ?

82. $4\frac{1}{2}$ times 7 is $\frac{1}{2}$ of what number ?

83. $\frac{2}{6}$ of 36 is $\frac{4}{5}$ of what number ?

84. $\frac{1}{3}$ of 36 is $\frac{12}{5}$ times what number ?

85. $\frac{2}{3}$ of A.'s age is 3 times B.'s age, and B is 9 years old; what is A.'s age ?

86. An individual, being asked the number of hours he labored each day, answered, $1\frac{1}{8}$ times the number of hours in a day is 3 times as many hours as I labor; how many hours did he labor each day ?

87. $\frac{3}{5}$ of 15 is $\frac{3}{7}$ of what number ?

88. $\frac{4}{7}$ of 21 is $1\frac{1}{5}$ times what number ?

89. $\frac{2}{3}$ of 24 is $1\frac{3}{5}$ times what number ?

90. Wright is 16 years old, and $1\frac{3}{4}$ times his age is $1\frac{2}{5}$ times Charles's age. How old is Charles ?

LESSON XXXVIII.

LESSONS XXXIV, XXXV, AND XXXVI COMBINED.

☞ REMARK.—*Pupils must exercise their judgment to employ the shortest of the methods given in Lessons XXXIV and XXXVI, for multiplying and dividing fractions.*

1. If 3 barrels of flour cost $\$13\frac{3}{4}$, what will 6 cost?

ANALYSIS.—$\$13\frac{3}{4}$ equals $\$\frac{55}{4}$. If 3 barrels of flour cost $\$\frac{55}{4}$, 6 barrels, which are 2 times 3 barrels, will cost 2 times $\$\frac{55}{4}$, which are $\$\frac{55}{2}$, or $\$27\frac{1}{2}$.

2. If 5 pounds of opium cost $\$27\frac{1}{2}$, what will 20 cost?
3. If 3 caps cost $\$17\frac{3}{8}$, what will 24 caps cost?
4. How many apples will pay for 9 oranges, if 8 apples are worth $12\frac{4}{5}$ oranges?

ANALYSIS.—$12\frac{4}{5}$ oranges equal $\frac{64}{5}$ oranges. If $\frac{64}{5}$ oranges are worth 8 apples, $\frac{1}{5}$ of an orange is worth $\frac{1}{64}$ of 8 apples, which is $\frac{8}{64}$, or $\frac{1}{8}$ of an apple; and $\frac{5}{5}$, or 1 orange, are worth 5 times $\frac{1}{8}$, or $\frac{5}{8}$ of an apple, and 9 oranges are worth 9 times $\frac{5}{8}$, which are $\frac{45}{8}$, or $5\frac{5}{8}$ apples.

REMARK.—In solving questions in Proportion, never seek the value of a unit of the denomination like the answer.

5. How many chestnuts will pay for 9 walnuts, if 7 chestnuts are worth $10\frac{2}{7}$ walnuts?
6. If 8 barrels of flour cost $\$33\frac{3}{5}$, what will 20 bbl. cost?
7. If it require $9\frac{3}{8}$ yards of cloth to make 3 coats, how many yards will it require to make 8 coats?
8. If 10 men can perform a piece of work in $9\frac{3}{5}$ days, how long will it take 8 men to perform the same?
9. What will be the cost of 6 sheep, if 15 cost $\$10\frac{1}{2}$?
10. If 1 person, in 6 months, consume $10\frac{7}{26}$ bushels of wheat, how much will 13 persons consume?
11. If $9\frac{7}{9}$ cents will buy 4 peaches, what will be the cost of 9 peaches?
12. If $\$9\frac{1}{16}$ will pay for 5 weeks' board, how many dollars will pay for 8 weeks' board?
13. If 6 orifices fill a vessel in $3\frac{3}{5}$ hours, how many of the same size will be required to fill it in $\frac{1}{15}$ of an hour?

14. If 9 men can build a boat in $5\frac{3}{4}$ days, in how many days could 6 men build it ?
15. If 2 men in 4 days earn $12, how many dollars can 7 men earn in the same time ?
16. If I pay $17\frac{2}{3}$ cents for riding 4 miles, how much must I pay for riding 6 miles ?
17. What will 1 year's board cost, at $\$21\frac{2}{3}$ for 4 weeks ?
18. If 9 barrels of fish cost $\$54\frac{1}{3}$, what will 27 cost ?
19. How many dollars will 1 barrel of tobacco cost, if 17 barrels cost $51\frac{1}{2}$ Eagles ?
20. If 13 pounds of tea cost $10\frac{2}{5}$ dimes, what will 5 pounds cost ?
21. If $7\frac{2}{3}$ tons of hay keep 6 horses through the winter, how many tons will keep 9 horses the same time ?
22. A fox is 40 rods before a hound, and runs 3 rods to the hound 5; how many rods must the hound run to overtake the fox ? How far did the fox run ?
23. How many dollars will a man earn in 14 days, if he earn $\$3\frac{2}{7}$ in 4 days ?
24. A merchant bought 8 pieces of cloth, each piece containing 5 yards, for $\$32\frac{1}{2}$; how much did it cost a piece, and how much a yard ?
25. If in a certain time 6 horses eat $14\frac{3}{4}$ bushels of oats, how many bushels will 8 horses eat in the same time ?
26. A boy sold 3 lemons, at the rate of 6 for 8 cents; how much did he receive for them ?
27. A boy gave $4\frac{1}{2}$ cents for oranges, at the rate of 5 oranges for $7\frac{1}{2}$ cents; how many did he buy ?
28. If a piece of mahogany, weighing 9 pounds, is worth $\$2\frac{3}{4}$, what is the value of 12 pounds, at the same rate ?
29. If a pole 8 feet long cast a shadow $4\frac{4}{5}$ feet, what will be the length of the shadow of a pole which is 15 feet long, at the same time of day ?
30. At a certain time of day, a pole 5 feet long casts a shadow $7\frac{1}{2}$ feet; what is the length of that pole which at the same time casts a shadow $4\frac{1}{2}$ feet ?

31. If it require $21\frac{2}{3}$ worth of provisions to serve 8 men 2 days, how many dollars' worth will serve 5 men 4 days?

32. What is the length of a pole the shadow of which is 12 feet long, at the same time, a pole $2\frac{2}{3}$ feet in length casts a shadow 4 feet long?

LESSON XXXIX.

To *Reduce* a fraction to a fraction having a given denominator,

Multiply both numerator and denominator of the given fraction by a number that will cause the denominator to become the required denominator.

[See Lessons XXXIV and XXXVI; also, Stoddard's New Practical Arithmetic, page 96, **142**, Prop. 5.]

1. $\frac{3}{4}$ is how many eighths?

ANALYSIS 1ST.—There are $\frac{8}{8}$ in 1, and in $\frac{1}{4}$ there is $\frac{1}{4}$ of $\frac{8}{8}$, or $\frac{2}{8}$, and in $\frac{3}{4}$ there are 3 times $\frac{2}{8}$, or $\frac{6}{8}$.

ANALYSIS 2D.—$\frac{3}{4}$ equal $\frac{6}{8}$.

2. $\frac{1}{4}$ is how many eighths?
3. $\frac{1}{5}$ is how many tenths?
4. $\frac{1}{2}$ is how many sixths?
5. $\frac{1}{3}$ is how many sixths?
6. $\frac{1}{2}$ and $\frac{1}{3}$ are how many sixths?
7. $\frac{2}{3}$ are how many sixths?
8. $\frac{5}{6}$ are how many twelfths?
9. $\frac{2}{6}$ are how many twelfths?
10. $\frac{3}{6}$ are how many eighteenths?
11. $\frac{3}{5}$ are how many tenths?
12. $\frac{2}{5}$ are how many tenths?
13. $\frac{4}{5}$ are how many twentieths?
14. $\frac{1}{3}$ is how many tenths?
15. Harris gave $\frac{3}{5}$ of an orange to his sister; how many fifteenths did he give her?

16. How many sixteenths in $\frac{3}{8}$?

17. How many sixteenths in $\frac{5}{8}$?

18. How many sixteenths in $\frac{7}{8}$?

19. How many fourteenths in $\frac{3}{7}$?

20. How many fourteenths in $\frac{5}{7}$?

21. How many fourteenths in $\frac{4}{7}$?

22. How many ninths in $\frac{2}{3}$?

23. How many twentieths in $\frac{4}{8}$?

24. How many fortieths in $\frac{7}{8}$?

25. How many forty-ninths in $\frac{5}{7}$?

26. How many fifteenths in $\frac{2}{5}$?

27. A man gave $\frac{1}{5}$ of a bushel of potatoes to one poor woman, and $\frac{3}{10}$ of a bushel to another; what part of a bushel did he give to both?

28. How could you divide an apple so as to give $\frac{2}{3}$ of it to 1 boy, and $\frac{1}{4}$ of it to another?

29. $\frac{2}{5} + \frac{2}{3}$ are how many fifteenths?

ANALYSIS.—$\frac{2}{5}$ equals $\frac{6}{15}$, and $\frac{2}{3}$ equals $\frac{10}{15}$. $\frac{6}{15}$ and $\frac{10}{15}$ are $\frac{16}{15}$, or $1\frac{1}{15}$.

30. $\frac{3}{4} + \frac{2}{3}$ are how many twelfths?

31. $\frac{8}{9} + \frac{1}{2}$ are how many eighteenths?

32. $\frac{7}{8} + \frac{2}{3}$ are how many twenty-fourths?

33. What is the sum of $\frac{3}{7}$ and $\frac{3}{12}$?

34. What is the sum of $\frac{3}{4}$ and $\frac{3}{8}$?

35. What is the sum of $\frac{2}{7}$ and $\frac{3}{5}$?

36. What is the sum of $\frac{1}{3}$ and $\frac{3}{7}$?

37. What is the sum of $\frac{2}{3}$ and $\frac{3}{5}$?

38. What is the sum of $\frac{3}{7}$ and $\frac{5}{6}$?

39. What is the sum of $\frac{1}{2}$ and $\frac{2}{3}$?

40. What is the sum of $\frac{5}{9}$ and $\frac{2}{3}$?

41. What is the sum of $\frac{7}{9}$ and $\frac{1}{2}$?

42. What is the sum of $\frac{3}{8}$ and $\frac{4}{7}$?

43. What is the sum of $\frac{5}{8}$, $\frac{1}{2}$, and $\frac{2}{3}$?

44. What is the sum of $\frac{3}{4}$, $\frac{2}{3}$, and $\frac{5}{6}$?

45. From $\frac{3}{4}$ subtract $\frac{1}{2}$.
46. From $\frac{5}{6}$ subtract $\frac{2}{3}$.
47. From $\frac{5}{7}$ subtract $\frac{2}{3}$.
48. From $2\frac{1}{3}$ subtract $\frac{3}{7}$.
49. From 4 subtract $\frac{1}{2}$.
50. From 3 subtract $\frac{3}{5}$.
51. From 9 subtract $\frac{3}{7}$.
52. From 5 subtract $\frac{4}{5}$.
53. From 3 subtract $1\frac{1}{2}$.
54. From 9 subtract $2\frac{2}{3}$.
55. From 6 subtract $3\frac{3}{4}$.
56. $14 - 3\frac{1}{2}$ are how many?
57. $7 - 2\frac{5}{6}$ are how many?
58. $9 - 3\frac{4}{7}$ are how many?
59. $10 - 3\frac{5}{8}$ are how many?
60. $12 - 3\frac{4}{9}$ are how many?
61. $13 - 7\frac{2}{11}$ are how many?
62. $9\frac{2}{3} - 4\frac{1}{2}$ are how many?
63. $7\frac{2}{3} - 5\frac{2}{7}$ are how many?
64. $3\frac{1}{2} - 1\frac{4}{5}$ are how many?
65. $4\frac{3}{7} - 1\frac{5}{6}$ are how many?
66. $5\frac{3}{5} - 2\frac{2}{3}$ are how many?
67. $9\frac{3}{5} - 7\frac{2}{3}$ are how many?
68. $2\frac{3}{4} + 3\frac{1}{4} - \frac{3}{5} =$ how many?
69. $4\frac{2}{3} + 5\frac{3}{4} - 2\frac{3}{5} =$ how many?
70. $3\frac{5}{6} + 4\frac{1}{6} - \frac{7}{8} =$ how many?
71. $9\frac{2}{3} + 3\frac{3}{4} - 3 =$ how many?
72. $\frac{2}{3} + \frac{3}{4} - \frac{7}{8} =$ how many?
73. $\frac{5}{6} + \frac{1}{3} + \frac{5}{7} - \frac{1}{2} =$ how many?
74. $2\frac{1}{8}$ are how many times $\frac{2}{16}$?
75. $3\frac{1}{5}$ are how many times $\frac{4}{10}$?
76. $\frac{9}{13}$ are how many times $\frac{3}{26}$?
77. $\frac{1}{2}$ are how many times $\frac{6}{36}$?

78. $\frac{4}{7}$ are how many times $\frac{2}{14}$?

79. $1\frac{1}{8}$ are how many times $\frac{3}{16}$?

80. $\frac{7}{9}$ are how many times $\frac{2}{18}$?

81. $\frac{5}{13}$ are how many times $\frac{5}{26}$?

82. $8\frac{2}{3}$ are how many times $2\frac{1}{6}$?

83. $10\frac{1}{5}$ are how many times $\frac{9}{15}$?

84. $12\frac{1}{2}$ are how many times $1\frac{1}{4}$?

85. $\frac{1}{6}+\frac{1}{5}+\frac{1}{4}$ is how much less than a whole one ?

86. $\frac{1}{7}+\frac{2}{6}+\frac{2}{5}$ is how much less than a whole one ?

87. A lady gave $\frac{1}{2}$ of all her money for a dress, and $\frac{2}{5}$ of it for a shawl; what part of her money remained ?

88. $\frac{1}{3}$ of an army was killed, and $\frac{1}{4}$ taken prisoners; what part of the army escaped ?

89. $\frac{3}{8}$ of an army was killed, $\frac{2}{5}$ taken prisoners, and 500 escaped; how many were there in the army ?

90. $\frac{2}{3}$ of the length of a pole is in the ground, $\frac{1}{5}$ of it in the water, and 12 feet in the air; what is the length of the pole ?

91. A market woman sold $\frac{2}{5}$ of all her oranges to one man, $\frac{1}{3}$ of them to another, and then had only 9 remaining; how many had she at first, and how many did she sell to each ?

92. A man, after spending $\frac{2}{3}$ of his fortune, found that $20 was $\frac{2}{9}$ of what he had remaining; what was his fortune ?

93. A hawk caught $\frac{2}{5}$ of Euphemia's chickens, a cat killed $\frac{1}{3}$ of them, $\frac{1}{4}$ of them died, and she had 13 remaining; how many had she at first, and how many were destroyed by the hawk and cat respectively ?

94. Said A. to B., if to my age you add its $\frac{1}{2}$ and its $\frac{2}{5}$, the sum will be 38; how old was he ?

95. A. is 40 years old, and $\frac{3}{4}$ of his age is $\frac{3}{5}$ of twice as much as his wife's age; how old is his wife ?

LESSON XL.

To *Reduce* a fraction to its lowest terms,

Divide each term of the given fraction by the largest number that is contained in each of them without a remainder.

NOTE.—A fraction is expressed in its lowest terms when no number greater than 1 will divide each of them without a remainder.

1. Reduce $\frac{4}{8}$ to its lowest terms.
2. Reduce $\frac{8}{9}$ to its lowest terms.
3. Reduce $\frac{3}{6}$ to its lowest terms.
4. Reduce $\frac{5}{15}$ to its lowest terms.
5. Reduce $\frac{4}{12}$ to its lowest terms.
6. Reduce $\frac{12}{24}$ to its lowest terms.
7. Reduce $\frac{15}{25}$ to its lowest terms.
8. Reduce $\frac{35}{75}$ to its lowest terms.
9. Reduce $\frac{36}{48}$ to its lowest terms.
10. Reduce $\frac{25}{35}$ to its lowest terms.
11. Reduce $\frac{36}{60}$ to its lowest terms.
12. Reduce $\frac{75}{100}$ to its lowest terms.
13. Reduce $\frac{32}{56}$ to its lowest terms.
14. Reduce $\frac{50}{70}$ to its lowest terms.
15. Reduce $\frac{12}{18}$ to its lowest terms.
16. Why does the value of the fraction remain the same, when you divide both *numerator* and *denominator* by the same number?
17. When you multiply both numerator and denominator by the same number, why does it not change the value of the fraction?
18. Reduce 4 times $\frac{6}{12}$ to its lowest terms.
19. Reduce 7 times $\frac{4}{14}$ to its lowest terms.
20. Reduce 8 times $2\frac{1}{4}$ to its lowest terms.
21. Reduce 6 times $\frac{5}{24}$ to its lowest terms.
22. Reduce 4 times $\frac{5}{15}$ to its lowest terms.
23. Reduce 12 times $\frac{3}{18}$ to its lowest terms.

L E S S O N   X L I.

1. If you cut an apple into 2 equal pieces, what part of 1 apple will 1 of these pieces be called?

2. If you cut $\frac{1}{2}$ of an apple into two equal pieces, what part of a whole apple will 1 of these pieces be called?

3. If Alice has $\frac{1}{3}$ of a lemon, and gives $\frac{1}{2}$ of it to Ann, what part of a lemon will Ann receive?

    ANALYSIS.—$\frac{1}{3}$ equals $\frac{2}{6}$. Therefore, Ann receives $\frac{1}{2}$ of $\frac{2}{6}$ of a lemon, which is $\frac{1}{6}$ of a lemon.

4. George, having $\frac{1}{2}$ of a melon, gave $\frac{1}{2}$ of it to Marcus; what part of a melon did Marcus receive?

5. Crary had $\frac{1}{4}$ of a dollar, and gave $\frac{1}{2}$ of it to Joshua; what part of a dollar did Joshua receive?

6. Robert had $\frac{1}{5}$ of a dollar, and gave $\frac{1}{2}$ of it for a cake; how many cents did the cake cost him?

7. Margaret had $\frac{1}{6}$ of a pound of candy, and Mary had $\frac{1}{2}$ as much; how much had Mary?

8. Jane had $\frac{1}{5}$ of a pound of sugar, and Ann $\frac{1}{3}$ as much; how much had Ann?

9. A boy bought $\frac{1}{5}$ of a quart of chestnuts, and gave $\frac{1}{3}$ of them to his sister; what part of a quart did she receive?

10. A man owned $\frac{1}{8}$ of a share in a bank, and sold $\frac{1}{2}$ of it; what part of a share had he remaining?

11. B. owned $\frac{1}{8}$ of a ship, and sold $\frac{1}{4}$ of his share; what part of a whole ship did he sell?

12. What is $\frac{1}{3}$ of $\frac{1}{5}$?

    ANALYSIS 1ST.—$\frac{1}{5}$ equals $\frac{3}{15}$, $\frac{1}{3}$ of $\frac{3}{15}$ is $\frac{1}{15}$; therefore, $\frac{1}{3}$ of $\frac{1}{5}$ equals $\frac{1}{15}$.

    ANALYSIS 2D.—$\frac{1}{3}$ of $\frac{1}{5}$ is $\frac{1}{15}$.

13. What is $\frac{1}{6}$ of $\frac{1}{2}$?

14. What is $\frac{1}{3}$ of $\frac{1}{7}$?

15. What is $\frac{1}{4}$ of $\frac{1}{6}$?

16. What is $\frac{1}{8}$ of $\frac{1}{9}$?

17. What is $\frac{1}{7}$ of $\frac{1}{5}$?

18. What is $\frac{1}{7}$ of $\frac{1}{4}$?

19. What is $\frac{1}{2}$ of $\frac{1}{6}$?

20. What is $\frac{1}{11}$ of $\frac{1}{3}$?

21. A kite up in the air fell $\frac{1}{8}$ of its height, it then arose $\frac{1}{6}$ of its distance from the ground; what part of the whole distance was it above the ground?

22. Homer is $\frac{1}{6}$ as old as his father, and Nelson is $\frac{1}{9}$ as old as Homer; what part of the father's age is Nelson's age?

23. A man, owning $\frac{1}{4}$ of a barrel of fish, accommodated his neighbor with $\frac{1}{7}$ of what he owned: how much had he remaining?

24. A man, having $\frac{1}{2}$ of an Eagle, gave $\frac{1}{5}$ of it to B, and B gave $\frac{1}{10}$ of what he had to C; how many cents had each?

25. Elizabeth had $\frac{2}{3}$ of a pie, and gave $\frac{1}{3}$ of her piece to Harriet; how much of the pie did Harriet receive?

ANALYSIS 1ST.—$\frac{1}{3}$ of $\frac{1}{3}$ is $\frac{1}{9}$. If $\frac{1}{3}$ of $\frac{1}{3}$ is $\frac{1}{9}$, $\frac{1}{3}$ of $\frac{2}{3}$ is twice $\frac{1}{9}$, which are $\frac{2}{9}$. Therefore, Harriet had $\frac{2}{9}$ of a pie.

ANALYSIS 2D.—$\frac{1}{3}$ of $\frac{2}{3}$ is $\frac{2}{9}$.

26. What is $\frac{1}{2}$ of $\frac{3}{5}$?

27. What is $\frac{1}{4}$ of $\frac{2}{3}$?

28. What is $\frac{1}{5}$ of $\frac{2}{3}$?

29. What is $\frac{1}{6}$ of $\frac{3}{4}$?

30. What is $\frac{1}{7}$ of $\frac{3}{8}$?

31. What is $\frac{1}{8}$ of $\frac{3}{5}$?

32. What is $\frac{1}{9}$ of $\frac{2}{7}$?

33. What is $\frac{1}{9}$ of $\frac{4}{5}$?

34. What is $\frac{1}{8}$ of $\frac{3}{6}$?

35. What is $\frac{1}{18}$ of $\frac{3}{5}$?

36. What is $\frac{1}{4}$ of $\frac{9}{13}$?

37. What is $\frac{1}{6}$ of $\frac{3}{4}$?

38. What is $\frac{1}{7}$ of $\frac{6}{7}$.

39. What is $\frac{1}{2}$ of $1\frac{3}{4}$?

40. What is $\frac{2}{3}$ of $1\frac{1}{4}$?

5

41.   What is $\frac{3}{4}$ of $\frac{1}{5}$?

42.   What is $\frac{3}{4}$ of $\frac{1}{7}$?

43.   What is $\frac{3}{5}$ of $\frac{1}{9}$?

44.   What is $\frac{5}{6}$ of $\frac{1}{7}$?

45.   What is $\frac{3}{7}$ of $\frac{1}{5}$?

46.   What is $\frac{3}{8}$ of $\frac{1}{9}$?

47.   What is $\frac{7}{8}$ of $\frac{1}{4}$?

48.   What is $\frac{5}{7}$ of $\frac{1}{8}$?

49.   What is $\frac{4}{6}$ of $\frac{1}{6}$?

50.   What is $\frac{3}{7}$ of $\frac{1}{4}$?

51.   What is $\frac{2}{3}$ of $\frac{3}{5}$?

52.   What is $\frac{3}{4}$ of $\frac{2}{5}$?

53.   What is $\frac{3}{5}$ of $\frac{2}{7}$?

54.   What is $\frac{3}{4}$ of $\frac{2}{8}$?

55.   What is $\frac{3}{4}$ of $\frac{3}{5}$?

56.   What is $\frac{3}{8}$ of $\frac{5}{7}$?

57.   What is $\frac{3}{5}$ of $\frac{6}{9}$?

58.   What is $\frac{4}{7}$ of $\frac{8}{9}$?

59.   What is $\frac{5}{7}$ of $\frac{6}{7}$?

60.   What is $\frac{8}{13}$ of $\frac{2}{3}$?

61.   What part of 1 is $\frac{2}{7}$ of $\frac{1}{6}$?

62.   What part of 2 is $\frac{4}{6}$ of $\frac{1}{3}$?

ANALYSIS 1ST.—$\frac{4}{6}$ of $\frac{1}{3}$ is $\frac{4}{18}$. $\frac{4}{18}$ is $\frac{4}{18}$ of 1, and is, therefore, $\frac{1}{2}$ of $\frac{4}{18}$, which is $\frac{1}{9}$ of 2.

ANALYSIS 2D.—$\frac{4}{6}$ of $\frac{1}{3}$ is $\frac{4}{18}$. 1 is $\frac{1}{2}$ of 2. If 1 is $\frac{1}{2}$ of 2, $\frac{1}{18}$ is $\frac{1}{2}$ of $\frac{1}{18}$, or $\frac{1}{16}$ of 2, and $\frac{4}{18}$ are 2 times $\frac{1}{16}$, or $\frac{1}{9}$ of 2.

64.   What part of 2 is $\frac{1}{4}$ of $1\frac{4}{5}$?

65.   What part of 2 is $\frac{3}{4}$ of $\frac{1}{3}$?

66.   What part of 3 is $\frac{1}{2}$ of $\frac{1}{2}$?

67.   What part of 4 is $\frac{1}{3}$ of $1\frac{4}{5}$?

68.   What part of 5 is $\frac{1}{5}$ of $\frac{2}{3}$?

69.   What part of 9 is $\frac{3}{7}$ of $1\frac{1}{3}$?

70.   What part of 2 is $\frac{2}{8}$ of $\frac{3}{4}$?

71. What part of 2 is $\frac{3}{7}$ of $2\frac{1}{2}$?
72. What part of 4 is $\frac{3}{5}$ of $\frac{8}{10}$?
73. What part of 6 is $\frac{2}{7}$ of $\frac{5}{6}$?
74. What part of 3 is $\frac{1}{2}$ of $4\frac{2}{5}$?
.75. What part of 4 is $\frac{5}{6}$ of $12\frac{5}{6}$?
76. What part of 7 is $\frac{2}{3}$ of $10\frac{3}{4}$?

77. Anthony had $\frac{1}{3}$ of $\frac{3}{4}$ of a pound of cinnamon; what part of a pound had he?

78. Albert had $\frac{1}{7}$ of $\frac{2}{3}$ of a quart of strawberries; how many strawberries had he, provided 1 quart contained 42 strawberries?

79. Abner gave $\frac{1}{4}$ of $\frac{6}{7}$ of a melon to his brother; what part of a melon had he remaining?

80. Matilda bought $\frac{5}{6}$ of a quart of milk for tea, and spilled $\frac{1}{3}$ of it; what part of a quart had she remaining?

81. Edwin picked $\frac{7}{8}$ of a pailful of blackberries, and on his way home spilled $\frac{1}{4}$ of them; what part of a pailful had he remaining?

82. A merchant bought $\frac{4}{5}$ of a hogshead of molasses, and lost $\frac{1}{5}$ of it by leakage; what part of a hogshead had he remaining?

83. Miriam had $\frac{2}{3}$ of a pound of candy, and gave $\frac{3}{4}$ of it to Augusta; what part of a pound did she give Augusta?

84. Elisha found $\frac{3}{4}$, and gave $\frac{2}{5}$ of it to Ephraim; what part of a dollar had Elisha remaining?

85. Andrew bought $\frac{1}{2}$ of a pound of maple-sugar, and gave $\frac{3}{4}$ of it to Walter; what part of a pound did Walter receive?

86. Jacob, having a pine-apple, gave $\frac{2}{3}$ of $\frac{6}{7}$ of it to the one that could tell how much that would be; what part of it had Jacob remaining?

87. James gave $\frac{2}{3}$ of $\frac{3}{4}$ of a dime for a top; how many cents did the top cost him?

88. Robert gave $\frac{3}{4}$ of a dollar for a cap; how many cents did the cap cost him?

89. Mary gave $\frac{7}{8}$ of $1\frac{2}{3}$ dimes for a comb; how many cents did the comb cost her?

90. Clarinda gave $\frac{2}{3}$ of 6 dimes for a pair of gloves; how many cents did the gloves cost?

91. A man, having $4\frac{2}{3}$ barrels of flour, sold $\frac{2}{3}$ of it; how much remained unsold?

92. A man gave $\frac{2}{3}$ of $3\frac{1}{3}$ for a silver pencil; what was the cost of the pencil?

93. Jane worked $8\frac{1}{4}$ hours in a day, and Delilah worked only $\frac{4}{5}$ of as many; how many hours in a day did Delilah work?

94. B. gave $32\frac{2}{3}$ for a cow, which was $\frac{2}{3}$ as much as A. gave for his; how much more did A.'s cow cost than B.'s?

95. Darius is $18\frac{3}{4}$ years old, which is $\frac{3}{4}$ of Daniel's age; how old is Daniel?

96. If 1 yd. of cloth cost $5\frac{3}{4}$, what will $\frac{2}{3}$ yd. cost?

97. If 4 yd. of cloth cost $9\frac{1}{2}$, what will $\frac{4}{5}$ yd. cost?

98. If 5 bbl. of beer cost $18\frac{1}{2}$, what will $\frac{1}{2}$ bbl. cost?

99. If $\frac{2}{3}$ of an apple cost $\frac{3}{4}$ of a cent, what will 1 apple cost?

100. If $\frac{1}{4}$ of an orange cost $1\frac{1}{3}$ cents, what will $\frac{5}{16}$ of an orange cost?

101. If 4 pounds of rice cost $6\frac{1}{4}$ dimes, how many cents will $1\frac{2}{3}$ pounds cost?

LESSON XLII.

1. If 4 bbl. of flour cost $14\frac{2}{5}$, what will $\frac{5}{6}$ of a bbl. cost?

2. If 3 bu. of pears cost $5\frac{1}{4}$, what will $1\frac{1}{6}$ bu. cost?

3. If $2\frac{1}{2}$ bushels of apples cost $6\frac{1}{4}$ dimes, how many cents will $\frac{4}{5}$ of a bushel cost?

4. If $\frac{3}{4}$ of an apple cost $\frac{2}{3}$ of a cent, what will 1 apple cost?

5. If $\frac{3}{8}$ of an orange cost $\frac{3}{4}$ of a cent, what will $\frac{3}{4}$ of an orange cost?

6. If $2\frac{3}{4}$ yd. of silk cost $\$3\frac{5}{8}$, what will $5\frac{1}{2}$ yd. cost?

7. If $5\frac{2}{5}$ yd. of satin cost $\$5\frac{4}{10}$, what will 2 yd. cost?

8. If in $3\frac{4}{7}$ hours, A. can do a piece of work, how long will it take him to do a piece $1\frac{2}{3}$ times as large?

9. $\frac{3}{4}$ of A.'s age is $\frac{2}{3}$ of B.'s; and $\frac{3}{4}$ of B.'s age is $\frac{2}{5}$ of C.'s age. How old are A. and B. respectively, provided C. is 81 years old?

10. Bought $3\frac{2}{3}$ boxes of goods, at $\$6\frac{8}{11}$ a box; how many sheep, at $2 each, will pay for them?

NOTE.—The expression, = ?, is read, *equals how many.*

11. $7 \div \frac{2}{5} = ?$ Or, 7 divided by $\frac{2}{5}$ equals how many?

ANALYSIS.—$\frac{2}{5}$ is contained in 1, $\frac{5}{2}$ times, and in 7, 7 times $\frac{5}{2}$, which are $\frac{35}{2}$, or $17\frac{1}{2}$ times.

12.	$2 \div \frac{2}{3} = ?$		15.	$6 \div \frac{5}{8} = ?$
13.	$3 \div \frac{3}{4} = ?$		16.	$9 \div \frac{9}{9} = ?$
14.	$5 \div \frac{4}{7} = ?$		17.	$12 \div \frac{8}{9} = ?$

18. How many times $\frac{2}{5}$ is $\frac{3}{4}$?

ANALYSIS 1st.—$\frac{2}{5}$ equals $\frac{8}{20}$, and $\frac{3}{4}$ equals $\frac{15}{20}$. $\frac{8}{20}$ is contained in $\frac{15}{20}$, $\frac{15}{8}$, or $1\frac{7}{8}$ times.

ANALYSIS 2D.—1 is contained in $\frac{3}{4}$, $\frac{3}{4}$ times. If 1 is contained in $\frac{3}{4}$, $\frac{3}{4}$ times, $\frac{1}{5}$ is contained in $\frac{3}{4}$, 5 times $\frac{3}{4}$ times, which are $\frac{15}{4}$ times, and $\frac{2}{5}$ is contained in it, $\frac{1}{2}$ of $\frac{15}{4}$ times, which is $\frac{15}{8}$, or $1\frac{7}{8}$ times.

19. How many times $\frac{2}{3}$ is $\frac{3}{4}$?

20.	$\frac{7}{8} \div \frac{2}{3} = ?$		29.	$5\frac{5}{9} \div \frac{5}{9} = ?$
21.	$\frac{6}{7} \div \frac{3}{5} = ?$		30.	$\frac{3}{10} \div \frac{2}{5} = ?$
22.	$\frac{3}{16} \div \frac{3}{8} = ?$		31.	$\frac{12}{14} \div \frac{4}{7} = ?$
23.	$1\frac{1}{3} \div \frac{2}{3} = ?$		32.	$3\frac{1}{4} \div \frac{5}{6} = ?$
24.	$\frac{4}{14} \div \frac{2}{7} = ?$		33.	$4\frac{1}{2} \div \frac{7}{8} = ?$
25.	$\frac{7}{8} \div \frac{2}{3} = ?$		34.	$2\frac{1}{4} \div 1\frac{1}{2} = ?$
26.	$3\frac{3}{7} \div \frac{4}{5} = ?$		35.	$1\frac{3}{5} \div 2\frac{2}{3} = ?$
27.	$2\frac{2}{3} \div \frac{3}{4} = ?$		36.	$5\frac{1}{5} \div 3\frac{1}{4} = ?$
28.	$2\frac{2}{5} \div \frac{2}{7} = ?$		37.	$5\frac{2}{3} \div 4\frac{1}{3} = ?$

38. A farmer sold a quantity of rye for $96, which was only $\frac{4}{5}$ of its value; how much did he lose?

39. A man sold a cow for $1\frac{2}{3}$ times what she cost him, and by so doing gained $6; how much did the cow cost him?

40. A merchant sold a quantity of goods for $\frac{13}{14}$ of what they cost him, and by so doing lost $15; how much did the goods cost?

41. A farmer, having lost 12 sheep, had only $\frac{7}{9}$ of his flock remaining; how many sheep had he left?

42. An individual being asked how many geese he had, answered, if to $\frac{5}{6}$ of his flock 24 geese were added, the number would equal $1\frac{4}{7}$ times his original flock; how many geese had he?

43. If $\frac{3}{4}$ of a yard of cloth cost $\$\frac{3}{4}$, what will $\frac{2}{5}$ of a yard cost?

44. A boy, being asked his age, said, that $8\frac{1}{4}$ years equaled $\frac{3}{4}$ of twice his age; how old was he?

45. If $\frac{3}{5}$ of my steamboat fare was $7\frac{1}{2}$, what was $\frac{7}{8}$ of it?

46. What will $\frac{5}{6}$ of a barrel of flour cost, if $\frac{4}{7}$ of a barrel cost $2\frac{1}{4}$?

47. What will $\frac{2}{3}$ of an orange cost, if $\frac{5}{7}$ of an orange cost $2\frac{1}{2}$ cents?

48. How many yards of cloth will be required to make a coat, if $1\frac{2}{3}$ yards will make $\frac{2}{3}$ of a coat?

49. $\frac{3}{4}$ of 2 are how many times $\frac{3}{2}$?

50. $\frac{3}{16}$ of 8 are how many times $\frac{1}{2}$?

51. $3 \div \frac{3}{5} \times 7 = ?$

52. $\frac{2}{5} \div \frac{3}{4} \times 8 = ?$

53. $\frac{2}{6} \times 6 \div \frac{3}{4} \times 12 = ?$

54. $\frac{3}{5} \times 2 \div \frac{2}{5} \times 7 = ?$

55. If $\frac{2}{5}$ of 3 yards of cloth cost $1\frac{1}{5}$, what will $\frac{3}{5}$ of 7 yards cost?

56. If $\frac{4}{7}$ of 6 yards of cloth cost $2\frac{3}{7}$, how much will $\frac{3}{5}$ of 7 yards cost?

57. If $\frac{2}{3}$ of $\frac{3}{4}$ of a barrel of flour cost $1\frac{2}{3}$, what will $\frac{1}{2}$ of $\frac{2}{3}$ of $\frac{3}{4}$ of a barrel cost?

LESSON XLIII.

1. 12 is $\frac{3}{4}$ of what number?

ANALYSIS.—If $\frac{3}{4}$ of some number is 12, $\frac{1}{4}$ of that number is $\frac{1}{3}$ of 12, or 4; and $\frac{4}{4}$, which is that number, are 4 times 4, which are 16. Therefore, 12 is $\frac{3}{4}$ of 16.

2. 15 is $\frac{3}{5}$ of what number?
3. 18 is $\frac{2}{7}$ of what number?
4. 20 is $\frac{4}{5}$ of what number?
5. 26 is $\frac{2}{3}$ of what number?
6. 25 is $\frac{5}{7}$ of what number?
7. 30 is $\frac{5}{8}$ of what number?
8. 32 is $\frac{8}{9}$ of what number?
9. 36 is $\frac{6}{7}$ of what number?
10. 36 is $\frac{6}{11}$ of what number?
11. 36 is $\frac{9}{2}$ of what number?
12. 24 is $\frac{6}{7}$ of what number?
13. 9 is $\frac{3}{7}$ of what number?
14. 12 is $\frac{4}{5}$ of what number?
15. 38 is $\frac{2}{3}$ of what number?
16. 16 is $\frac{4}{6}$ of what number?
17. 16 is $\frac{4}{7}$ of what number?
18. 16 is $\frac{2}{5}$ of what number?
19. 40 is $\frac{5}{8}$ of what number?
20. 40 is $\frac{8}{9}$ of what number?

ANALYSIS.—40 is $\frac{8}{9}$ of 45.

REMARK.—When pupils are familiar with the analysis of these questions, the intermediate steps may be omitted, as in the above analysis.

21. 72 is $\frac{8}{9}$ of what number?
22. 72 is $\frac{9}{10}$ of what number?
23. 12 is $\frac{2}{3}$ of how many times 2?
24. 16 is $\frac{4}{9}$ of how many times 3?
25. 18 is $\frac{2}{5}$ of how many times 9?

26. 32 is $\frac{4}{7}$ of how many times 4?
27. 46 is $\frac{2}{3}$ of how many times 23?   -
28. 48 is $\frac{4}{5}$ of how many times 5?
29. 48 is $\frac{4}{7}$ of how many times 4?
30. 36 is $\frac{6}{7}$ of how many times 2?
31. 30 is $\frac{5}{6}$ of how many times $\frac{1}{2}$ of 12?
32. 30 is $\frac{6}{7}$ of how many times $\frac{1}{2}$ of 10?
33. 16 is $\frac{4}{9}$ of how many times $\frac{2}{3}$ of 9?
34. 16 is $\frac{4}{15}$ of how many times $\frac{3}{4}$ of 16?
35. 24 is $\frac{8}{9}$ of how many times $\frac{2}{3}$ of 12?
36. 25 is $\frac{5}{9}$ of how many times $\frac{1}{3}$ of 9?
37. 35 is $\frac{5}{12}$ of how many times $\frac{2}{3}$ of 9?
38. 40 is $\frac{5}{8}$ of how many times $\frac{4}{5}$ of 10?
39. 48 is $\frac{6}{10}$ of how many times $\frac{4}{5}$ of 25?
40. 96 is $\frac{2}{3}$ of how many times $\frac{3}{4}$ of 16?

LESSON XLIV.

1. $\frac{2}{3}$ of 6 is $\frac{2}{5}$ of what number?
2. $\frac{2}{5}$ of 10 is $\frac{2}{3}$ of what number?
3. $\frac{3}{4}$ of 8 is $\frac{2}{7}$ of what number?
4. $\frac{2}{7}$ of 21 is $\frac{3}{4}$ of what number?
5. $\frac{1}{5}$ of 15 is $\frac{3}{10}$ of what number?
6. $\frac{3}{10}$ of 40 is $\frac{4}{5}$ of what number?
7. $\frac{3}{9}$ of 27 is $\frac{3}{5}$ of what number?
8. $\frac{8}{9}$ of 27 is $\frac{8}{13}$ of what number?
9. $\frac{5}{9}$ of 81 is $\frac{9}{10}$ of what number?
10. $\frac{2}{7}$ of 49 is $\frac{6}{11}$ of what number?
11. $\frac{2}{3}$ of 12 is $\frac{1}{2}$ of how many times 2?
12. $\frac{3}{4}$ of 16 is $\frac{2}{3}$ of how many times 2?
13. $\frac{4}{5}$ of 10 is $\frac{2}{7}$ of how many times 4?
14. $\frac{1}{4}$ of 16 is $\frac{2}{15}$ of how many times 6?

15. $\frac{4}{5}$ of 15 is $\frac{4}{7}$ of how many times 6?

16. $\frac{2}{5}$ of 20 is $\frac{2}{3}$ of how many times 3?

17. $\frac{5}{6}$ of 12 is $\frac{2}{9}$ of how many times 5?

18. $\frac{3}{4}$ of 20 is $\frac{4}{9}$ of how many times 3?

19. $\frac{7}{6}$ of 36 is $\frac{2\frac{1}{2}}{10}$ of how many times 4?

20. $\frac{8}{9}$ of 72 is $\frac{2}{3}$ of how many times 12?

SOLUTION.—$\frac{8}{9}$ of 72 is 64. 64 is $\frac{2}{3}$ of 96. 96 is 8 times 12.

21. $\frac{9}{8}$ of 96 is $\frac{2}{5}$ of how many times 90?

22. $\frac{7}{9}$ of 117 is $\frac{7}{8}$ of how many times 4?

23. $\frac{5}{7}$ of 56 is $\frac{5}{6}$ of how many times 8?

24. $\frac{5}{6}$ of 60 is $\frac{2}{3}$ of how many times 5?

25. $\frac{2}{3}$ of 36 is $\frac{2}{5}$ of how many times 12?

26. $\frac{3}{4}$ of 72 is $\frac{9}{8}$ of how many times 5?

27. $\frac{7}{8}$ of 40 is $\frac{5}{12}$ of how many times 21?

28. $\frac{5}{8}$ of 32 is $\frac{4}{9}$ of how many times 9?

29. $\frac{2}{3}$ of 15 is $\frac{5}{7}$ of how many times 2?

30. $\frac{2}{5}$ of 15 is $\frac{1}{3}$ of how many times 9?

31. $\frac{5}{8}$ of 24 is $\frac{5}{4}$ of how many times 3?

32. $\frac{7}{9}$ of 45 is $\frac{5}{6}$ of how many times 3?

33. $\frac{5}{7}$ of 14 is $\frac{2}{9}$ of how many times 5?

34. $\frac{7}{9}$ of 18 is $\frac{2}{9}$ of how many times 7?

35. $\frac{9}{10}$ of 40 is $\frac{3}{8}$ of how many times 6?

36. $\frac{4}{5}$ of 45 is $\frac{6}{11}$ of how many times 3?

37. $\frac{6}{7}$ of 35 is $\frac{5}{9}$ of how many times 2?

38. $\frac{8}{9}$ of 81 is $\frac{4}{5}$ of how many times 9?

39. $\frac{2}{3}$ of 5 is $\frac{5}{7}$ of how many times 7?

40. $\frac{3}{4}$ of 7 is $\frac{3}{5}$ of how many times 3?

41. B.'s horse cost $60, and $\frac{4}{5}$ of the cost of the horse is $\frac{1}{4}$ of 2 times the value of his wagon; what is the value of his wagon?

42. A coat cost $30, and $\frac{4}{5}$ of the cost of the coat is $\frac{2}{7}$ of 8 times the price of a hat; the price of the hat is required.

43. If a cow cost $30, and $\frac{2}{3}$ of this is $\frac{4}{7}$ of 10 times the price of a sheep, what is the price of a sheep?

44. A.'s farm is worth $1200, and $\frac{1}{4}$ of its value is $\frac{3}{8}$ of 10 times the value of its yearly productions; what is the value of the yearly productions?

45. The articles contained in a certain store cost $500, and $\frac{3}{10}$ of their cost is $\frac{2}{5}$ of 3 times the amount paid for the silks; how much was the cost of the silks and of the other articles respectively?

46. A.'s wedding clothes cost $180, and $\frac{2}{3}$ of the cost of his clothes is $\frac{2}{5}$ of 6 times the cost of his bride's wedding dress; how much was the cost of her dress?

47. The insurance of a ship amounted to $800, and $\frac{1}{4}$ of that is $\frac{1}{20}$ of 2 times the value of the cargo; what is the value of the cargo?

48. A.'s house cost $1400, and $\frac{1}{4}$ of its cost is $3\frac{1}{2}$ times $\frac{1}{2}$ of the cost of the furniture contained in it; what was the cost of the furniture?

49. Provided a house was worth $1200, and $\frac{3}{4}$ of its value was $\frac{2}{3}$ of $\frac{1}{2}$ times the value of the farm on which it stood; what was the value of the farm?

50. If a sleigh cost $100, what would be the cost of a wagon, if $\frac{2}{3}$ of the cost of the sleigh was $\frac{2}{11}$ of twice the cost of a wagon?

51. Before the war Lambert's property was worth $2500, and $\frac{4}{5}$ of its value then is $3\frac{1}{2}$ times $\frac{1}{2}$ of its value after the war; what was its last value?

Distances on the Railroad between Albany and Buffalo.

52. The distance from Albany to Schenectady is 16 miles, and $\frac{3}{4}$ of this distance is $\frac{2}{3}$ of $\frac{1}{6}$ of the distance from Albany to Rome; what is the distance to Rome?

53. Fort-Plain is 56 miles from Albany, and $\frac{5}{7}$ of this distance is $1\frac{2}{3}$ times $\frac{1}{10}$ of the distance from Albany to Rochester; what is the distance to Rochester?

54. Waterloo is 192 miles from Albany, and $\frac{5}{6}$ of this

distance is $1\frac{3}{8}$ times the distance from Albany to Utica, and 3 miles more; what is the distance to Utica?

55. Buffalo is 325 miles from Albany, and $\frac{3}{5}$ of this distance is $7\frac{1}{2}$ times $\frac{1}{11}$ of the distance to Batavia, and 5 miles more; what is the distance from Albany to Batavia?

Distances on the Railroad between Albany and Boston.

56. Boston is 200 miles from Albany, and $\frac{3}{5}$ of this distance is $1\frac{3}{5}$ times $\frac{1}{2}$ of the distance to West Springfield; what is the distance to West Springfield?

57. From Albany to the State line is 38 miles, and $1\frac{1}{2}$ times this distance is $4\frac{3}{4}$ times $\frac{1}{9}$ of the distance to Wilbraham; what is the distance to Wilbraham?

58. Kinderhook is 16 miles from Albany, and $\frac{3}{4}$ of this distance is $\frac{2}{3}$ times $\frac{1}{3}$ of the distance to Dalton; what is the distance to Dalton?

59. Brighton is 195 miles from Albany, and $\frac{2}{5}$ of this distance is $\frac{1}{4}$ of 2 times the distance to Worcester; what is the distance to Worcester?

60. Grafton is 162 miles from Albany, and $\frac{5}{9}$ of this distance is $\frac{1}{3}$ of 3 times the distance to Westfield, less 2 miles; how far is it to Westfield?

LESSON XLV.

1. $\frac{2}{3}$ of 9 is $\frac{3}{5}$ of how many times $\frac{1}{5}$ of 25?
2. $\frac{3}{4}$ of 16 is $\frac{1}{3}$ of how many times $\frac{1}{7}$ of 21?
3. $\frac{4}{5}$ of 40 is $\frac{2}{3}$ of how many times $\frac{1}{2}$ of 16?
4. $\frac{7}{8}$ of 80 is $\frac{2}{5}$ of how many times $\frac{1}{3}$ of 21?
5. $\frac{3}{4}$ of 36 is $\frac{1}{3}$ of how many times $\frac{3}{4}$ of 12?
6. $\frac{8}{9}$ of 45 is $\frac{2}{7}$ of how many times $\frac{5}{7}$ of 14?
7. $\frac{5}{6}$ of 30 is $\frac{5}{3}$ of how many times $\frac{1}{2}$ of 10?
8. $\frac{11}{12}$ of 48 is $\frac{1}{3}$ of how many times $\frac{2}{7}$ of 7?

9. $\frac{4}{5}$ of 35 is $\frac{2}{9}$ of how many times $\frac{3}{4}$ of 8 ?

   ANALYSIS.—$\frac{4}{5}$ of 45 is 36. 36 is $\frac{2}{9}$ of 162. $\frac{3}{4}$ of 8 is 6. 162 is 27 times 6.

10. $\frac{5}{7}$ of 35 is $\frac{1}{2}$ of how many times $\frac{3}{7}$ of 11$\frac{2}{3}$?
11. $\frac{7}{9}$ of 54 is $\frac{2}{3}$ of how many times $\frac{2}{3}$ of 10$\frac{1}{2}$?
12. $\frac{3}{5}$ of 25 is $\frac{3}{4}$ of how many times $\frac{2}{5}$ of 10 ?
13. $\frac{4}{7}$ of 28 is $\frac{2}{5}$ of how many times $\frac{4}{5}$ of 25 ?
14. $\frac{4}{9}$ of 18 is $\frac{2}{9}$ of how many times $\frac{3}{4}$ of 12 ?
15. $\frac{5}{6}$ of 36 is $\frac{3}{8}$ of how many times $\frac{5}{6}$ of 12 ?
16. $\frac{4}{9}$ of 54 is $\frac{4}{8}$ of how many times $\frac{3}{4}$ of 16 ?
17. $\frac{7}{8}$ of 32 is $\frac{2}{3}$ of how many times $\frac{2}{3}$ of 9 ?
18. $\frac{8}{9}$ of 108 is $\frac{3}{8}$ of how many times $\frac{2}{5}$ of $\frac{2}{5}$ of 15 ?
19. $\frac{3}{4}$ of 40 is $\frac{5}{12}$ of how many times $\frac{1}{2}$ of $\frac{4}{5}$ of 20 ?
20. $\frac{4}{5}$ of 20 is $\frac{2}{5}$ of how many times $\frac{2}{5}$ of $\frac{3}{4}$ of 12 ?

LESSON XLVI.

1. If 1 horse eat $\frac{1}{4}$ of a bushel of oats in 1 day, how many horses will eat a bushel in the same time ?
2. If the wages of 8 weeks amount to $48, what will the wages of 2$\frac{3}{4}$ weeks amount to ?
3. A ship's crew of 12 men have provision for 5 months; how many months will it last 5 men ?
4. A man gained $14 by selling a watch for 1$\frac{2}{5}$ times what it cost him ; how much did it cost ?
5. There is a pole, $\frac{7}{8}$ of its length is under water, and 9 feet out ; how long is the pole ?
6. A pole is standing in the water, so that 15 feet is above the water, which is $\frac{3}{7}$ of the whole length of the pole ; how long is the pole ?
7. If $\frac{3}{4}$ be 2, what will 2 be ? Or, if $\frac{3}{4}$ of an apple cost 2 cents, how much will 2 apples cost ?
8. If 8 horses in 1 day eat 4 bushels of oats, in how many days can 1 horse eat 1 bushel ?

9. If 3 horses in 1 day eat $1\frac{1}{4}$ bushels of oats, how many bushels can 1 horse eat in 4 days?

10. If 1 horse in 2 days eat 6 bushels of corn, how many bushels will 4 horses eat in 3 days?

11. If 4 horses eat 16 bushels of grain in 2 days, how many bushels will 3 horses eat in 12 days?

12. How many tons of hay will 3 horses consume in 4 days, if 4 horses in $\frac{1}{2}$ of a day consume $\frac{4}{5}$ of a ton?

13. How many hundred-weight of hay can 3 horses consume in 25 days, if 2 horses in $\frac{1}{4}$ of a day consume $\frac{7}{400}$ of a hundred-weight?

14. In how many days can 4 men cut 16 cords of wood, if 1 man in 1 day cut $\frac{1}{3}$ of a cord?

15. How many men will be required to earn 20 dimes in 4 days, if 4 men in $2\frac{1}{4}$ days earn 11 dimes?

16. If it require 6 days for 2 men to lay 36 rods of wall, how many men can in $\frac{1}{2}$ of the time build 72 rods of similar wall?

17. If in 4 days 3 men accomplish a certain piece of work, how many men will be required to perform a piece of work 4 times as large in 2 days?

18. If 4 men in 8 days perform a certain piece of work, how many men will be required to accomplish 3 times as much work in $\frac{3}{4}$ of a day?

19. If 1 horse eat 1 bushel of oats in 4 days, in how many days would 6 horses eat 48 bushels?

20. If $\frac{2}{3}$ of 6 be 3, what will $\frac{1}{4}$ of 40 be?

21. If 3 be $\frac{2}{3}$ of 6, what will $\frac{1}{4}$ of 40 be?

22. If 2 men in $\frac{1}{4}$ of a day earn $\frac{5}{12}$ of a dollar, in how many days can 3 men earn $\frac{3}{4}$ of a dollar?

23. If it require $\frac{1}{8}$ of a bushel of oats to feed 4 horses $\frac{1}{3}$ of a day, how many horses would it require to consume 9 bushels in $\frac{2}{3}$ of a day?

☞ SUGGESTION.—*Review unless the pupils thoroughly understand the preceding Lessons.*

The study of the following "Arithmetical and Algebraic Problems" may be omitted until the class has learned the Lessons in Interest, commencing on page 141.

LESSON XLVII.

ARITHMETICAL AND ALGEBRAIC PROBLEMS.

1. 24 is $\frac{2}{5}$ of twice as much as a cask of wine cost; what did the wine cost?

2. Bought 30 barrels of flour, and $\frac{4}{5}$ of the number of barrels equaled $\frac{1}{6}$ as many dollars as they all cost; what did 1 barrel cost?

3. 35 is $\frac{5}{8}$ of how many times $\frac{2}{8}$ of 4?

4. A farmer, being asked how many sheep he had, answered, that 160 was $\frac{2}{5}$ of 10 times his number; how many sheep had he?

5. Mr. B., being asked the value of his horse, said, $54 is $\frac{6}{11}$ of 3 times its value; what is the value of his horse?

6. 72 is $\frac{8}{9}$ of how many times $\frac{3}{4}$ of 12?

7. 36 is $\frac{3}{4}$ of how many times $\frac{2}{3}$ of 12?

8. 48 is $\frac{2}{3}$ of how many times $\frac{1}{2}$ of 18?

9. 56 is $\frac{8}{9}$ of how many times $\frac{7}{8}$ of 8?

10. 60 is $\frac{3}{5}$ of how many times $\frac{5}{8}$ of 16?

11. 84 is $\frac{12}{5}$ of how many times $\frac{1}{5}$ of 25?

12. A. spent $60, which was $\frac{5}{6}$ of 4 times as much as he was worth; how much was he worth?

13. B. sold 9 sheep, which was $\frac{3}{10}$ times $\frac{1}{2}$ of his whole flock; how many sheep had he remaining?

14. D., at a game of cards, lost $20, which was $\frac{4}{7}$ times $\frac{5}{8}$ of all his money; how much had he?

15. C. found $45, which was $\frac{5}{6}$ of 3 times as much as he already had; how much more did he find than he had at first?

16. A boy lost 9 marbles, which was $\frac{3}{8}$ of twice the number he had at first; how many had he left?

17. A boy gave away 8 apples, which was $\frac{2}{7}$ of twice as many as he had left; how many had he at first?

18. 12 is $\frac{3}{5}$ times $\frac{3}{7}$ of what number?

19. 36 is $\frac{3}{11}$ times $\frac{3}{5}$ of how many times $\frac{3}{4}$ of $13\frac{1}{3}$?

20. Jeremiah is 18 years old, and his age is $\frac{3}{4}$ times $\frac{3}{5}$ of his father's age; how old is his father?

21. Mary gave 6 cents for a comb, which was $\frac{2}{5}$ times $\frac{1}{3}$ of all her money; how many cents had she?

22. Martha gave 8 cents for a pine-apple, which was $\frac{4}{7}$ times $\frac{2}{5}$ of all her money; how many apples could she have bought with the money she had remaining, at 2 cents each?

23. Henry had 20 marbles, which was $\frac{2}{3}$ of twice as many as Harry had; how many had Harry?

24. Margaret is 16 years old, and her age is $\frac{2}{5}$ of 3 times Martha's age; how old is Martha?

25. $\frac{3}{4}$ is $\frac{2}{5}$ of twice as much as what number?

26. A man bought a horse for $60, which was $\frac{3}{7}$ of twice as much as he sold it for; how much did he gain by the bargain?

27. A horse was sold for $40, which was $\frac{4}{5}$ times $\frac{5}{6}$ of what it was worth; what was the value of the horse?

28. A man when he was married, was 20 years of age, which was $\frac{5}{6}$ times $\frac{3}{7}$ of the age of his wife; how old was she?

29. Shepherd was worth $160, which was $\frac{4}{7}$ times $\frac{1}{10}$ of his father's fortune; required the father's fortune.

30. A. and B. were playing cards, B. lost $14, which was $\frac{7}{10}$ times $\frac{2}{3}$ as much as A. then had; and when they commenced, $\frac{5}{6}$ of A.'s money equaled $\frac{4}{7}$ of B.'s. How much had each when they began to play?

31. A. and B. were playing cards, A. lost $20, which was $\frac{5}{14}$ of the number of dollars B. then had more than A.; provided this sum was $1\frac{1}{6}$ times as much as A. had at first, how much had each when they began to play?

LESSON XLVIII.

1. A boy, after spending $\frac{3}{5}$ of all his money, found that 16 cents was all he had remaining; how much had he at first?

ANALYSIS 1ST.—Let $\frac{5}{5}$ equal the money he had at first. Then, after spending $\frac{3}{5}$ of it, he had remaining $\frac{5}{5}-\frac{3}{5}$, which is $\frac{2}{5}$. This, by the condition of the question, is 16 cents. If $\frac{2}{5}$ of the money he had at first is 16 cents, $\frac{1}{5}$ of it is $\frac{1}{2}$ of 16 cents, which is 8 cents; and $\frac{5}{5}$, or what he had at first, are 5 times 8, or 40 cents.

ANALYSIS 2D.—He spent $\frac{3}{5}$ of his money; therefore, he had remaining $\frac{2}{5}$ of it, which equals 16 cents. If $\frac{2}{5}$ of his money is 16 cents, he must have had 40 cents.

2. Ruth, after losing $\frac{2}{3}$ of all her roses, had only 3 remaining; how many had she at first?
3. Jane gave $\frac{3}{5}$ of all her flowers to Ann, and had 4 remaining; how many did she give to Ann?
4. George, after eating $\frac{9}{13}$ of all his oranges, had only 8 oranges remaining; how many had he at first?
5. A boy expended $\frac{1}{9}$ of his money for a pie, $\frac{2}{9}$ for a ball, $\frac{3}{9}$ for a top, and had 6 cents remaining; how many cents had he at first?
6. In a certain school $\frac{1}{2}$ of the scholars study grammar, $\frac{1}{3}$ study arithmetic, and the remainder, which is 10, study geography; how many scholars in all, and how many attending to each study?
7. A third part of an army was killed, $\frac{1}{4}$ part taken prisoners, and 300 escaped; how many were there in the army?
8. If from my age you subtract $\frac{1}{2}$ and $\frac{2}{5}$ of my age, the remainder is 2 years; how old am I?
9. Mr. B., being asked how many pigeons he caught, said, that if to $\frac{5}{7}$ of the number 36 were added, the sum would equal twice the number; how many did he catch?

10. If to $\frac{3}{4}$ of the cost of B.'s horse you add \$100, the sum will be twice the cost of the horse; what was the oost of the horse?

11. A gentleman, after spending $\frac{3}{7}$ of his fortune and $\frac{1}{3}$ of the remainder, had \$2400 remaining; what was his fortune?

12. A gambler lost $\frac{3}{4}$ of all his money, and the next night he won $\frac{2}{3}$ as much as he lost the night before; he then had \$90. How much had he at first?

13. John had stolen from him $\frac{5}{7}$ of his money; the thief was not caught until he had spent $\frac{5}{7}$ of all he had stolen: the remainder, which was \$40 less than John had remaining, was given back; how much money had John at first?

14. A traveler had stolen from him $\frac{5}{7}$ of his money; the thief was not caught until he had spent $\frac{1}{3}$ of what he had stolen; the remainder, \$100, was given back; how much had he at first?

15. If to $\frac{1}{2}$ of the cost of A.'s watch you add \$10, the sum will be \$21; what was the cost of his watch?

16. If to $\frac{3}{5}$ of B.'s age, you add 15 years, the sum would be 39 years; how old is B.?

17. A drover, being asked how many sheep he had, said, if to $\frac{1}{3}$ of my flock you add the number $9\frac{1}{2}$, the sum will be $99\frac{1}{2}$; how many sheep had he?

18. $\frac{5}{7}$ of the length of a pole is in the water, and 12 feet in the air; how long is the pole?

19. If to $\frac{7}{8}$ of A.'s age you add 16 years, the sum will be $1\frac{1}{8}$ times his age; how old is he?

20. A man, being asked how many pigeons he caught, replied, if to $\frac{3}{4}$ of the number I caught you add 20, the sum would lack 4 of being equal to $1\frac{1}{3}$ times the number; how many did he catch?

LESSON XLIX.

1. Divide the number 36 into two parts, which shall be to each other as 7 to 2.

ANALYSIS.—Since the two parts are to be to each other as 7 to 2, we must divide 36 into 7 + 2, which are 9 equal parts; and 7 of the parts will be one of the numbers, and 2 of them the other. $\frac{1}{9}$ of 36 is 4, and $\frac{7}{9}$ are 7 times 4, which are 28 (the first number), and $\frac{2}{9}$ are 2 times 4, which are 8 (the other number). Or, in the latter part of the solution, say: $\frac{7}{9}$ of 36 is 28, the first number; and $\frac{2}{9}$ of 36 is 8, the other number.

2. 2 men hired a pasture for $72; one put in 7 horses, and the other 2 horses; what ought each to pay?

3. A. and B. hired a pasture for $14; A. put in 4 cows, and B. put in 3 cows; what ought each to pay?

4. A. and B. bought a lottery ticket for $5; A. paid $3, and B. paid $2. They drew a prize for $60; what was each one's share?

5. Two men bought 40 mules: the first paid $5 as often as the other paid $3. How many mules ought each to receive?

6. Mary and Elizabeth went to school 80 days, and as often as Mary went 3 days, Elizabeth went 5 days; how many days did each attend school?

7. Reuben had 7 cents, and Blake 4 cents; they paid all their money for 22 apples; how many apples ought each to receive?

8. Three men bought a lottery ticket for $12; the first paid $2, the second $7, and the third $3. They drew a prize of $240; what was each man's share?

9. Three men hired a pasture for $24; the first put in 2 horses, the second put in 3 horses, and the third put in 4 horses; how much ought each to pay?

10. A man, failing in business, was able to pay only $\frac{3}{4}$ of his debts; how much will that man receive to whom he owes $90?

11. A man, meeting an equal number of poor women and boys, gave to each woman 7 dimes, and to each

boy 2 dimes : and to them all he gave $9 ; how many women and boys were there respectively ?

12. Two men bought a barrel of fish for $9 ; the first paid $4, the second $5 ; what part of the barrel belongs to each ?

13. A farmer gave 35 bushels of rye to two neighbors; to the first he gave 1 bushel as often as to the other ⅔ of a bushel ; how many bushels did each receive ?

14. Three men hired a pasture for $36 : the first put in 3 horses, the second 2 horses, and the third 4 horses ; how much ought each to pay ?

15. Two men hired a pasture for $60 : the first put in 4 horses for 2 weeks, and the second put in 3 horses for 4 weeks ; how much ought each to pay ?

16. Three men hired a pasture for $15 : the first put in 4 sheep for 5 weeks, the second put in 8 sheep for 5 weeks, and the third put in 10 sheep for 9 weeks. How much ought each to pay ?

17. Two men entered into partnership ; the first put in $40 for 10 months, and the second put in $80 for 5 months ; they gained $95 : what was each man's share of the gain ?

18. A. and B. agreed to cut a field of wheat for $20 ; A. sent 5 men for 4 days, and B. sent 3 men for 10 days : how much ought each to receive ?

19. Divide $56 between A. and B., giving to A. $1 as often as to B. ⅔ of a dollar.

20. A. and B. hired a pasture for $24 ; A. put in 4 sheep for 10 weeks, and B. put in 2 horses for 10 weeks ; what ought each to pay, provided 2 sheep in 1 week eat as much as a horse in the same time ?

21. Simpson, Domer, and Eyer enter into a joint speculation by which they clear $460. Simpson claims to have furnished ¾ ; Domer, ⅔ ; and Eyer, ½ of the entire capital. How much, according to these calculations, ought each to receive ?

NOTE.—¾, ⅔, and ½ are to each other as 9, 8, and 6.

LESSON L.

1. Mr. B. had 4 apples more than A., and together they had 14; how many had each?

ANALYSIS.—By a condition of the question, B.'s number is equal to A.'s+4 apples; to which add A.'s number, and we have 2 times A.'s number+4=14. Therefore, 2 times A.'s number equals 14—4, or 10; and once his number equals ¼ of 10, or 5 apples. And B.'s number is 5+14=9 apples.

2. Heman has 6 books more than Handford, and both have 26; how many has each?

3. Robert has 7 marbles more than Richard, and both have 35; how many has each?

4. Mary has 4 roses more than Martha, and both have 24; how many has each?

5. Alice has 7 pins more than Abner, and both have 29; how many has each?

6. ⅔ of ¾ is ⅛ of what number?

7. The sum of two numbers is 36, and their difference is 16; what are the two numbers?

8. A boy bought ⅔ of a melon for 8⅓ cents; how much is that for 1 melon?

9. Homer and Hannah each bought an equal number of peaches; on their way home Hannah had 4 more given to her, then together they had 24: how many did each buy?

10. Two boys had each an equal number of blocks; one lost 4; and together they then had only 12 remaining: how many had each at first?

11. A wagon was sold for $17⅔, which was ⅔ as much as it cost; what did it cost?

12. Hiram had twice as many strawberries as Eugene, and both had 18 pints; how many had each?

13. Ida had 6 cents more than twice as many as Ira, and both had 36; how many had each?

14. Susan had ⅛ as many cents as Sarah; Sarah lost

10 : then together they had 50 ; how many had each at first ?

15. Thomas was returning from market with twice as many eggs as Timothy; Thomas broke 4 of his, and Timothy 6 of his ; they then had 50 eggs remaining. How many had each at first ?

16. ⅔ of a number +14=44; what is that number?

17. A boy being asked his age, replied, 3 times my age —7 years are 23 years; how old was he ?

18. Mr. A., being asked how much money he had, replied, twice what I have + $60, is four times $400; how much money had A. ?

19. Two boys have 49 marbles, but the first has 7 the most; how many has each ?

20. A man bought a sheep, a cow, and a horse for $70; the cow cost $10 more than the sheep, and the horse cost $20 more than the cow. What was the cost of each ?

21. A man bought a melon for 18⅘ cents, which was only ⅗ as much as his dinner cost ; what was the cost of his dinner ?

22. A gentleman bought a watch and chain for $80 ; the chain cost ⅓ as much as the watch; what was the cost of each?

23. A farmer bought a plow, a harness, and a horse for $58; for the harness he gave $6 more than for the plow, and for the horse $34 more than for the harness. How much did he give for each ?

24. A boy bought twice as many oranges as lemons, and on his way home ate 4 oranges and gave 6 away ; and was surprised to find he had only 14 oranges remaining. How many of each kind did he buy ?

25. 5 times a certain number — 12 is 48 ; what is that number ?

26. ¾ of a certain number — 5 is 40 ; what is that number ?

27. A boy, being asked his age, replied, 11 years are 7 years more than $\frac{2}{3}$ of my age; how old was he?

28. A boy, being asked how many sheep his father had, replied, 40 are 5 less than $\frac{3}{4}$ of his father's number; how many had he?

29. A boy bought 18 lemons; for $\frac{2}{3}$ of them he paid 3 cents for 2, and for the remainder he paid 3 cents each; for what must he sell them each to gain 10 cents on the whole?

30. James, John, and Joseph together have 96 peaches; James has 2 more than John, and Joseph has as many as both James and John: how many has each?

31. Henry bought 54 oranges; for $\frac{2}{3}$ of them he paid 2 cents for 3, and for the remainder, 3 cents for 2; and sold $\frac{1}{3}$ of them, at the rate of 2 cents for 3, and the remainder, at 3 cents for 2. How much did he gain by so doing?

LESSON LI.

1. If a man can do a certain piece of work in 12 days, what part of it can he do in 1 day?

2. If a man can drink a barrel of beer in 20 weeks, what part of it can he drink in 1 week?

3. If it require 9 hours to empty a vessel, what part of it can be emptied in 1 hour?

4. If a family consume a barrel of pork in 30 days, what part of a barrel do they daily consume?

5. If it require 19 days to perform a certain journey, what part of it can be performed in 1 day?

6. If A. can do a certain piece of work in 8 days, and B. could do the same in 12 days; what part of it can each do in a day?

7. If C. could mow a certain field in 4 days, and D. could do the same in 6 days, what part of it

could each do in a day? How much could they together do in a day?

8. If C. and D. can, in 1 day, mow $\frac{5}{12}$ of a field, how long would it take them to mow the whole field?

9. How many days would it take to perform a certain piece of work, if $\frac{3}{15}$ of it can be performed in 1 day?

10. If George can do a certain piece of work in 3 days, and Granvil in 6 days; how long will it take them together to do the work?

11. If James can eat a bushel of apples in 10 days, and Rud in 12 days; how long would 1 bushel last both?

12. A. can cut a field of wheat in 12 days, and B. can do the same in 20 days; how long would it take them to cut a field when they work together?

13. A merchant bought a hogshead of molasses for $20, 10 gallons of which leaked out; how must he sell the remainder a gallon to gain $6.50?

14. $\frac{2}{3}$ of a barrel of flour cost $4\frac{2}{3}$, what will $\frac{1}{4}$ of a barrel cost?

15. A. and B. can build a boat in 20 days, and with the assistance of C., they can build it in 8 days. How long would it take C. to build it alone?

16. A farmer and his son can do a piece of work in 6 days; the son can do the same in 27 days. How long would it take the father to do the work?

17. Three pipes, A, B, and C, can fill a cistern in 2 hours, A and B can fill it in 4 hours, and A and C can fill it in 3 hours. How long would it take each to fill it?

18. If a barrel of beer would last a man 35 days, and the man and his son 20 days; how long would it last the son alone?

19. A box of tea usually lasted a man and his wife 9 months; when the man was absent it would last the wife 12 months. How long would it have lasted the man alone?

20. A., B., and C. can build a boat in 20 days, A. and B., in 40 days, and A. and C., in 30 days. How long would it take each separately to build it?

21. Provided A. could drink a barrel of beer in 24 days, and B., in 36 days; how long would it take them together to drink a barrel, after $\frac{2}{8}$ of it had leaked out?

22. A market-woman bought 30 oranges, and had $\frac{1}{3}$ of them stolen; the remainder she sold at 3 cents each, and thereby gained $\frac{2}{3}$ of a cent on each orange bought. How much did they cost each?

23. A. can do a certain piece of work in $4\frac{1}{2}$ days, and A. and B. together in 3 days. After A. did $\frac{1}{4}$ of the work, B. did the remainder; how long did it take him?

24. If A. can do a certain piece of work in $\frac{2}{6}$ of a day, how much can he do in 1 day?

25. If a man can chop a cord of wood in $\frac{3}{7}$ of a day, how much can he chop in 1 day?

26. Isaac can make a pair of boots in $\frac{2}{8}$ of a day, and Ira in $\frac{2}{5}$ of a day; how many pair can both make in a day?

27. Samuel can cut a cord of wood in $\frac{3}{4}$ of a day, and Theodore in $\frac{2}{5}$ of a day; how long would it take them to cut a cord, when they worked together?

28. If $\frac{2}{8}$ of an apple cost $\frac{4}{7}$ of a cent, what will $\frac{5}{6}$ of an apple cost?

29. A. can mow 1 acre of grass in $\frac{2}{3}$ of a day, B., in $\frac{3}{4}$ of a day, and C., in $\frac{4}{5}$ of a day. How much more can A. and B. mow in a day than C.?

30. If a wolf can eat a sheep in $\frac{7}{8}$ of an hour, and a bear can eat it in $\frac{3}{4}$ of an hour, how long would it take them together to eat what remained of a sheep after the wolf had been eating $\frac{1}{2}$ of an hour?

NOTE.—The wolf in $\frac{1}{2}$ hour can eat $\frac{4}{7}$ of the sheep, and in 1 hour he can eat $\frac{8}{7}$ of a sheep. The bear in 1 hour can eat $\frac{4}{3}$ of a sheep. Therefore, in 1 hour both can eat $\frac{8}{7}+\frac{4}{3}=\frac{52}{21}$ of a sheep. Therefore, to eat $\frac{3}{7}-\frac{4}{7}=\frac{9}{21}$ of a sheep, it will require $\frac{9}{52}$ of an hour = 10 min. $23\frac{1}{13}$ sec.

LESSON LII.

1. Lewis, meeting some beggars, gave each of them 2 cents, and had 12 cents remaining; if he had given them 4 cents each, it would have taken all the money he had. How many beggars were there?

ANALYSIS.—By the last condition of the question, he gave each beggar 2 cents more than by the first, and to them all, 12 cents *more* than by the first condition. Therefore, there must have been as many beggars as 2 is contained times in 12, which are 6 beggars.

2. A boy gave to each of his playmates 3 cents, and had 24 cents remaining; if he had given them each 7 cents, it would have taken all the money he had. How many playmates had he?

3. Mary gave each of her playmates 5 apples; if she had given them each 7 apples, it would have taken 12 apples more. How many playmates had she?

4. A number of persons gave me 10 cents each; had they given me 12 cents each, it would have amounted to 20 cents more. How many persons were there?

5. $\frac{2}{5}$ of $100 is $\frac{3}{250}$ of $\frac{1}{4}$ times the salary of the President of the United States; what is his salary?

6. $40 is $\frac{2}{5}$ times $\frac{1}{16}$ of the salary of the Vice-President of the United States; what is his salary?

7. Divide 35 oranges between James and Joseph, so that James may have 15 more than Joseph.

8. A cask of wine was sold for $96, which was $\frac{2}{3}$ of twice as much as it cost; how much did it cost?

9. By selling a quantity of cotton for $560, I gained $\frac{3}{4}$ of what it cost; how much did it cost?

10. A. and B. are 187 miles apart, and are traveling towards each other, one at the rate of 8 miles an hour, and the other, 9 miles an hour; how many hours before they meet?

11. Agnes gave 2 dimes a yard for a piece of calico; had she given 3 dimes a yard, it would have cost 20 dimes more; how many yards did the piece contain?

12. A. was ordered to buy a certain number of oranges; if he bought those, at 2 cents each, he would have had no money left; if he bought those at 3 cents each, he would have wanted 10 cents more to have paid for them. How many oranges was he required to buy?

13. A lady wished to buy a certain number of yards of muslin; there were two kinds, some at 9 cents a yard, and some at 12 cents a yard. Had she taken that at 12 cents a yard, it would have cost 36 cents more than the other kind. How many yards did she wish to buy?

14. A boy being sent to market to buy a certain number of pounds of meat, found, if he bought beef at 5 cents a pound, he would have 39 cents remaining, but if he bought pork at 8 cents a pound, he would have only 6 cents remaining. For how much meat was he sent?

15. If 8 times a certain number is 36 more than 5 times the same number, what is that number?

16. A boy being asked his age, said, 4 times my age is 24 years more than 2 times my age; how old was he?

17. A boy, being asked how many sheets of paper he had, said, 4 times the number is 18 less than 7 times the number; how many sheets of paper had he?

18. A person, wishing to buy some butter, found, if he bought that which was 10 cents a pound, he would have 20 cents remaining; but if he bought that which was 12 cents a pound, he would lack 14 cents of having money enough to pay for it. How many pounds did he wish to buy?

19. A farmer, wishing to buy a certain number of sheep, found if he gave $2 a head, he would have $20 remaining; but if he gave $5 a head, he would lack $40 of having money enough to pay for them. How many sheep did he wish to buy?

20. A., B., and C., talking of their ages; says A. to B., I am 4 times as old as you; says B. to C., I am ½ as old as you; but says A. to C., I am 40 years older than you. Required the age of each.

LESSON LIII.

1. A laborer agreed to work 40 days upon this condition: that for every day he worked he should receive $2, and for every day he was idle he should pay $1 for his board. At the expiration of the time, he received $50. How many days did he work?

ANALYSIS.—If he had labored the whole time, he would have received 40 times $2, or $80. But he received only $50: he, therefore, lost by his idleness $80 − $50, or $30. For every day he was idle he lost $2 (his daily wages) + $1 (the cost of his board), which are $3. If in 1 day he lose $3, he will lose $1 in ⅓ of a day, and $30 in 30 times ⅓ of a day, or 10 days. Therefore, he was idle 10 days, and worked 40 − 10 days, or 30 days.

2. A man agreed to work 60 days on this condition: that for every day he worked he should receive $1½, and for every day he was idle he should pay $½ for his board. At the expiration of the time, he received $68. How many days did he work?

3. A man was hired for 80 days, on this condition: that for every day he worked he should receive 6 dimes, and for every day he was idle he should forfeit 4 dimes. At the expiration of the time, he received $40. How many days did he work?

4. How many times ⅔ of 12 is ⅞ of 48?

5. A. and B. bought a quantity of flour for $50; A. paid $1 as often as B. $\frac{2}{3}$ of a dollar; what part of the flour belongs to each?

6. A., B., and C. built a house, which cost $500, of which B. paid $100 more than A., and C. paid as much as A. and B. both; how much did each pay?

7. A merchant sold a quantity of cloth for $84, and thereby lost $\frac{2}{3}$ of what it cost; what did it cost?

8. $7\frac{1}{2}$ is $2\frac{1}{2}$ times $\frac{2}{3}$ of how many times $1\frac{1}{2}$?

9. A farmer having in his employ an equal number of men and boys, gave to each boy $4, to each man $8, and to them all, $84; how many men were there?

10. Two men hired a pasture for $35; one put in 3 cows, and the other 4 cows; how much ought each to pay?

11. A man sold an equal number of ducks and turkeys for 20 dimes; the ducks at 2 dimes each, and the turkeys at 3 dimes each. How many did he sell in all?

12. A farmer sold an equal number of ducks and turkeys; the ducks at 4 dimes each, the turkeys at 7 dimes each; and for the turkeys he received $3 more than for the ducks. How many of each did he sell?

13. There are two baskets, containing 37 apples; in one of which there are 17 more than in the other. How many apples in each?

14. Charles and Henry together have 49 marbles, and Charles has 7 more than twice as many as Henry. How many has each?

15. Philip has 20 apples more than Philo; and together they have 92; how many has each?

16. Three boys have 47 lemons; the first has 3 more than the second, and the second 7 more than the third. How many has each?

17. A boy was hired for 20 days, on this condition:

that for every day he labored he should receive 3 dimes, and for every day he was idle he should pay 2 dimes for his board. At the expiration of the time he received only $1. How many days was he idle?

18. A boy bought a whistle, a whip, and a drum for 70 cents. For the whip he gave twice as much as for the drum, and for the drum, twice as much as for the whistle; how much did he give for each?

19. The sum of three numbers is 54. The first is twice, and the third 3 times the second; what are those numbers?

20. Sarah's age is ⅔ of Susan's, and the sum of their ages is 24; what is the age of each?

21. ⅝ of an army were killed, ¾ of the remainder taken prisoners, and 400 escaped; how many were there in the army?

LESSON LIV.

1. A fishing rod, the length of which was 14 feet, was broken into two pieces. The shorter piece was ¾ of the length of the longer. What was the length of each piece?

> ANALYSIS.—¾ of the length of the longer piece, which is the length of the shorter, +¼, (the length of the longer) = ¼ of the length of the longer, which is the length of both, or 14 feet. If ¼ of the longer is 14 feet, ¼ is ¼ of 14 feet, which is 2 feet, and ¼ (which is the length of the longer) are 4 times 2 feet, or 8 feet. 14 − 8 = 6 feet, the length of the shorter piece.

2. A pole, the length of which is 20 feet, is in the air and water; ¾ of the length in the air equals the length in the water. What is the length in the air and water respectively?

3. If in 2 days a man traveled 160 miles, and ¾ of the distance he traveled the first day, equals the

distance he traveled the second day; how far did he travel each day?

4. B. and C. together have 40 marbles; how many has each, provided $\frac{2}{5}$ of B.'s number is equal to C.'s?

5. From New York City to Redhook is 100 miles, and $\frac{1}{8}$ of the distance from New York to Rhinebeck, equals the distance from Rhinebeck to Redhook. How far from Rhinebeck to New York, and how far from Rhinebeck to Redhook?

6. If a horse and a colt were worth $90, and the horse was worth $1\frac{1}{2}$ times as much as the colt, what was the value of each?

7. A boy paid 70 cents for a slate and a book; how much did he pay for each, provided the book cost $1\frac{1}{3}$ times as much as the slate?

8. If a traveler pay $1.20 for his breakfast and dinner, how much did he pay for each, provided his dinner cost $\frac{4}{5}$ as much as his breakfast?

9. A pole, the length of which is 67 feet, is in the air and water; $\frac{2}{3}$ of the length in the air + 7 feet equals the length in the water. Required the length in the air and in the water.

10. Divide the number 108 into two such parts, that $\frac{3}{4}$ of the first + 8 shall equal the second.

11. Divide the number 97 into two such parts, that $\frac{3}{4}$ of the first + 7 shall equal the second.

12. There is a fish the length of which is 18 feet; its tail is 4 feet, and $\frac{4}{5}$ of the length of the body equals the length of the head. What is the length of the head and body respectively?

13. There is a fish the weight of which is 11 pounds, and $\frac{1}{2}$ of the weight of the head + 8 pounds equals the weight of the body; what is the weight of each?

14. A ship-mast 51 feet in length, was broken off in a storm, and $\frac{2}{3}$ of the length broken off, equaled $\frac{3}{4}$ of the length remaining; how much was broken off. and how much remained?

15. A boy being asked how many apples and oranges he had, answered, in all I have 36, and $\frac{2}{5}$ of the number of apples equals $\frac{1}{2}$ of the number of oranges; how many of each kind had he?

16. $\frac{2}{3}$ of one number equals $\frac{3}{5}$ of another, and their sum is 57; what are the two numbers?

17. A farmer has 290 sheep in two different fields; and $\frac{3}{4}$ of the number in the first field, equals $\frac{2}{7}$ of the number in the second; how many are there in each field?

18. A market woman was requested to buy 33 fowls, consisting of two kinds; $\frac{1}{4}$ of the number of the first kind, was to equal $\frac{2}{3}$ of the number of the second; how many of each must she buy?

19. A person, being asked the time of day, said, the time past noon is $\frac{1}{7}$ of the time past midnight; what was the hour?

REMARK.—Since the time past noon is $\frac{1}{7}$ of the time past midnight, the time from midnight to noon, which is 12 hours, must be $\frac{6}{7}$ of the time past midnight.

20. A person, being asked the hour of the day, said, the time past noon is $\frac{1}{3}$ of the time past midnight; what was the hour?

21. A person, being asked the hour of the day, said, the time past noon is $\frac{1}{3}$ of the time from now to midnight; what is the hour?

ANALYSIS.—From "now" to midnight is $\frac{3}{3}$, and $\frac{1}{3}$ added (the time past noon) is $\frac{4}{3}$. Consequently, from noon to midnight (which is 12 hours) is $\frac{4}{3}$ of the time it lacked of being midnight; and $\frac{1}{3}$ of the time is $\frac{1}{4}$ of 12, which is 3 hours, the time past noon.

22. What is the time of day, provided $\frac{3}{4}$ of the time from now to midnight equals the time past noon?

23. A man, being asked the hour of the day, said, $\frac{2}{3}$ of the time past noon equals $\frac{2}{9}$ of the time from now to midnight; what was the time?

A pole, the length of which was 68 feet, was in the air and water; $\frac{3}{4}$ of the length in the air equaled $\frac{2}{3}$ of the length in the water. What was the length in the air and in the water respectively?

25. The sum of two numbers is 176, and $\frac{3}{4}$ of the first $+4$ equals $\frac{2}{3}$ of the second; required the numbers.

26. A person, being asked the time of day, said, $\frac{3}{5}$ of the time past midnight equals $\frac{3}{10}$ of the time from now to midnight again; what o'clock is it?

27. Provided the time past 10 o'clock, A. M., equals $\frac{3}{4}$ of the time to midnight; what o'clock is it?

28. Says A. to B., $\frac{3}{4}$ of my age + 4 years equals $\frac{2}{3}$ of yours, and the sum of our ages is 74 years; what is each of their ages?

29. A person, being asked the hour of the day, replied, $\frac{3}{5}$ of the time past noon equals $\frac{2}{9}$ of the time from now to midnight + $2\frac{2}{3}$ hours; what was the time?

30. A pole, the length of which is 78 feet, is in the air and water; $\frac{2}{8}$ of the length in the air + 12 feet, equals $1\frac{1}{2}$ times the length in the water. What is the length in the air and water respectively?

LESSON LV.

1. There is a fish the head of which is 4 inches long, and whose tail is as long as its head + $\frac{1}{2}$ of its body, and whose body is as long as its head and tail; what is the length of the fish?

ANALYSIS.—By a condition of the question, $\frac{1}{2}$ of the length of the body + 4 inches, is the length of the tail; to which add 4 inches (the length of the head), and we have $\frac{1}{2}$ of the length of the body + 8 inches = $\frac{2}{2}$, or the length of the body. Therefore, $\frac{2}{2} - \frac{1}{2}$, or $\frac{1}{2}$ of the length of the body, equals 8 inches; and $\frac{4}{2}$, or twice the length of the body, which is the length of the fish, equals 4 times 8 inches, or 32 inches.

2. The head of a fish is 6 inches long, its tail is as long as its head + $\frac{1}{2}$ of its body, and the body is

as long as the head and tail together; what is the length of the fish?

3. The head of a fish is 12 inches long, the tail is as long as the head + $\frac{1}{2}$ of the body, and the body is as long as the head and tail; what is the length of the fish?

4. The head of a fish weighs 10 pounds, the tail weighs as much as the head + $\frac{2}{3}$ as much as the body, and the body weighs as much as the head and tail; what is the weight of the tail?

5. $\frac{5}{7}$ of a certain number equals $\frac{2}{7}$ of the same number + 10; what is that number?

6. A boy, being asked his age, replied, $\frac{2}{3}$ of my age exceeded $\frac{2}{5}$ of my age by 4 years; how old was he?

7. James, being asked how many arithmetical questions he had answered correctly during the week, replied, $\frac{3}{4}$ of the number is 3 more than $\frac{3}{5}$ of the number; how many questions had he answered?

8. A farmer, after selling $\frac{2}{3}$ of $1\frac{1}{5}$ times as much grain as he had, had 80 bushels remaining; how much had he at first?

9. An individual, after spending $\frac{2}{3}$ of all his money, and $\frac{2}{3}$ of what then remained, had only $\$12\frac{2}{3}$ remaining; how much had he at first?

10. If $\frac{6}{7}$ of a ship be worth $\frac{4}{5}$ of her cargo, which is valued at 300 Eagles, what is the value of the ship?

11. Dick being asked how much money he had, said, its $\frac{1}{2}$ exceeded its $\frac{3}{8}$ by \$2; how much had he?

12. A tree, by falling, was broken into three pieces; the top part was 10 feet long, the bottom part was as long as the top + $\frac{3}{5}$ of the middle, and the middle part was as long as the other two; what was the length of the tree, and of each piece?

13. A man bought a hat, a coat, and a watch; the hat cost \$6, the watch cost as much as the hat + $\frac{3}{5}$ of the cost of the coat, and the coat cost as

much as the hat and watch; what was the cost of each, and of all?

14. $\frac{2}{3}$ of $\frac{3}{4}$ is how many times $\frac{2}{5}$ of $\frac{5}{16}$?

15. If $\frac{3}{4}$ of a ton of hay cost $\frac{2}{3}$ of an Eagle, how many dollars will $\frac{1}{3}$ of a ton cost?

16. A third and $\frac{1}{2}$ of a third of 10 is $\frac{5}{8}$ of what number?

17. If from a certain number you take its $\frac{1}{2}$ and its $\frac{1}{3}$, the remainder will be $13\frac{2}{3}$; what is that number?

18. After spending $\frac{2}{3}$ of my money, I earned $\frac{2}{3}$ as much as I spent, and then had only $20 less than what I had at first; how much had I at first?

19. The head of a fish is 8 inches long, the tail is as long as the head and $\frac{1}{2}$ of the body + 10 inches, and the body is as long as the head and tail; what is the length of the fish?

20. The head of a fish is 12 inches long; its tail is 10 inches longer than its head increased by $\frac{1}{2}$ the length of the body, and its body is 20 inches longer than its head and tail together; what is the length of the fish?

LESSON LVI.

1. James is 20 years old, and John is 4 years old; in how many years will James, who is now 5 times as old as John, be only twice as old?

> REMARK.—Four years ago James was 16, and in 16 years more he will be twice as old as John.

2. Sarah is 10 years old, and Sally is 4; in how many years will Sally be $\frac{1}{2}$ as old as Sarah?

3. Jacob is 40 years old, and Alfred is 2; in how many years will Alfred be $\frac{1}{2}$ as old as Jacob?

4. *If* a third of 6 be 3, what will $\frac{1}{4}$ of 20 be?

5. *If* 3 be a third of 6, what will $\frac{1}{4}$ of 20 be?

6. *If* $\frac{3}{4}$ of 12 be 10, what will $\frac{2}{3}$ of 10 be?

7. Divide the number 85 into two parts, that shall be to each other as $\frac{2}{3}$ to $\frac{3}{4}$.

8. When A. was married, he was 3 times the age of his wife, but 15 years after their marriage his age was only twice her age; how old was each when they were married?

   REMARK.—The conditions of the above question give the following:

9. Three times a certain number + 15 equals twice the same number + 30; what is that number, and what is 3 times the same number?

10. Once a certain number + 15 equals $\frac{2}{3}$ of the same number + 30; what is that number, and what is $\frac{1}{3}$ of the same number?

11. When I first met Mr. A., I was $\frac{1}{2}$ as old as he was, and in 12 years after that I was $\frac{3}{4}$ as old as he was; what was each of our ages when we first met?

12. There are two numbers, one of which is 4 times the other; but if to each 20 were added, one will be double the other; what are these numbers?

13. When B. was married he was 3 times as old as his wife; but after they had been married 60 years, $\frac{2}{3}$ of his age equaled hers; what was the age of each when they were married?

14. A hound takes 3 leaps to a fox 4, and 3 of the hound's leaps are equal to 6 of the fox's; how many leaps must the hound take to gain 1 on the fox?

15. If the hound takes $1\frac{1}{2}$ leaps to gain 1 on the fox, how many must he take to gain 20 on the fox?

16. A hare is 20 leaps before a hound, and takes 4 leaps to the hound 3; and 3 of the hound's leaps are equal to 6 of the hare's. How many leaps must the hound take to catch the hare?

17. A fox is 60 leaps before a hound, and takes 5 leaps to the hound 2; and 4 of the hound's leaps equals 12 of the fox's. How many leaps must the hound take to catch the fox?

18. Alfred is 60 steps before Silas, and takes 9 steps to Silas 6; and 3 of Silas's steps equals 7 of Alfred's. How many steps, at this rate, will each take before they will be together?

REMARK.—*A box of glass contains* 50 *square feet, as nearly as may be.*

19. How many panes of glass in a box, provided they are 6, by 8 inches?

REMARK.—Find the area of a pane of glass by reducing the inches to parts of a foot, and then multiply these parts together. The area of a pane 6 by 8 inches is $\frac{1}{3}$ of a square foot. The remainder may be solved as follows:

ANALYSIS.—If to make $\frac{1}{3}$ of a square foot it require 1 pane, to make $\frac{3}{3}$, or 1 square foot, it will require 3 times 1, or 3 panes; and to make 50 square feet (1 box), it will require 50 times 3 panes, which are 150 panes.

20. How many panes of glass in a box, provided they are 8, by 10 inches?
21. How many panes of glass in a box, provided they are 10, by 12 inches?
22. How many panes of glass in a box, provided they are 8, by 12 inches?
23. How many panes of glass in a box, provided they are 12, by 15 inches?

LESSON LVII.

1. What number is that, to which if its $\frac{1}{2}$ be added, the sum will be 15?
2. What number is that, to which if its $\frac{1}{2}$ be added, the sum will be 24?
3. What number is that, to which if its $\frac{1}{3}$ be added, the sum will be 40?
4. What number is that, to which if its $\frac{1}{4}$ be added, the sum will be 30?
5. What number is that, to which if its $\frac{3}{8}$ be added, the sum will be 88?

6. How old is that man, to whose age if you add its $\frac{1}{3}$ and its $\frac{2}{3}$, the sum will be 104 years?

7. What number is that, which being increased by its $\frac{1}{2}$, its $\frac{1}{3}$, and 18 more, will be doubled?

8. A man, being asked his age, said, my age increased by its $\frac{2}{3}$ and 20 more, is double my age. What was his age?

9. Suppose I buy a certain number of boxes of butter, at $2 a box, as many more at $4 a box, and sell them all at $3 a box; do I gain or lose, and how much?

10. A boy, being asked how many oranges he had, replied, if my number were increased by its $\frac{2}{3}$, its $\frac{3}{4}$, and 42 more, the sum would equal 3 times my number. How many had he?

11. Suppose I buy a certain number of melons; some at 10 cents each, and as many more at 40 cents each; and sell them all at 30 cents each; how much do I gain on each melon?

12. If by selling 1 apple I lose $\frac{3}{20}$ of a cent, how many apples, at this rate, must I sell to lose 6 cents?

13. A boy bought a certain number of lemons, at 2 cents each, as many more at 4 cents each; and sold them at the rate of 3 for 5 cents: did he gain or lose, and how much?

14. A woman bought a certain number of apples, at the rate of 2 for a cent, as many more at the rate of 3 for a cent; and sold them all at the rate of 5 for 2 cents, and by so doing, lost 4 cents. How many of each kind did she buy?

15. A woman bought a certain number of eggs, at the rate of 3 for a cent, as many more at 4 for a cent; and sold them out at the rate of 8 for 3 cents, and by so doing, gained 4 cents. How many eggs did she buy?

16. Three men agreed to share $510 in the proportion of $\frac{1}{2}$, $\frac{2}{3}$, and $\frac{1}{4}$; how much must each receive?

17. A.'s money is to B.'s as $\frac{1}{2}$ is to $\frac{1}{3}$; and they together have $100; how much has each?

18. The difference of two numbers is 15, which is $\frac{3}{4}$ of twice as much as the smaller number; what are these two numbers?

19. A merchant bought a number of yards of cloth, at the rate of 2 yards for $1, and as many more, at the rate of 5 yards for $1; and sold all the cloth at the rate of 10 yards for $3; and thereby lost $8. How many yards did he buy?

20. A man bought a certain number of melons, at the rate of 4 for $1, as many more at the rate of 10 for $1; and sold them all at the rate of 8 for $2, and thereby gained $6. How many melons did he buy?

LESSON LVIII.

1. Mary has twice as many apples as Sarah, and together they have 12; how many has each?

REMARK.— By the condition of the question, Mary has 2 apples as often as Sarah 1. Consequently, Mary must have $\frac{2}{3}$, and Sarah $\frac{1}{3}$ of the 12 apples.

2. Divide 18 into 2 such parts, that one shall be twice the other.

3. Divide 21 oranges between two boys, so that one may have twice as many as the other.

4. Franklin and Francis together have 15 quarts of nuts, but Franklin has twice as many as Francis; how many quarts has each?

5. Robert has twice as many cents as Harry, and together they have 24; how many has each?

6. Divide the number 27 into two parts, that shall be to each other as 1 to 2.

7. Harriet is twice as old as Ellen, and the sum of their ages is 30 years; what is the age of each?

8. A. and B. are 36 rods apart, and travel towards each other; how far will each travel before they meet, provided A. travels twice as fast as B. ?

9. What number must be added to twice itself, that the sum may be 57 ?

10. A., after spending ¼ of all his money, and ⅔ of the remainder less $4, had only $14 remaining; how much had he at first ?

11. Divide the number 48 into two such parts, that one shall be ⅔ of the other.

12. In a certain school, there are 3 times as many boys as girls, and in all there are 52 pupils; how many boys and how many girls in the school ?

13. James and Jackson together have 45 marbles, but James has only ¼ as many as Jackson; how many has each ?

14. A man and his son together earned $280 in a year; how much did each earn, provided the boy earned only ⅓ as much as his father ?

15. A boy bought a melon and a citron for $1; how much did each cost, provided the melon was in value only ¼ as much as the citron ?

16. A man bought a horse and a saddle for $120; the saddle cost only ⅕ as much as the horse; what was the cost of each ?

17. A man, being asked the cost of his oxen, said, my oxen and wagon together cost $240, and the oxen cost twice as much as the wagon; what was the cost of each ?

18. A man paid for a sheep, a hog, and a cow, $42; for the hog, he gave twice as much as for the sheep, and for the cow, 3 times as much as for the sheep. How much did he give for each ?

19. A man and his two sons earned $560 in 1 year; the father earned twice as much as his elder son, and the elder son earned twice as much as the younger son. How much did each earn ?

20. A., B., and C. together, in 1 day, can dig 105 bushels of potatoes; A. can dig $\frac{1}{2}$ as much as B., and B., $\frac{1}{2}$ as much as C. How many bushels can each dig in a day?

21. A man bought 3 pieces of cloth for $160; the first piece cost only $\frac{1}{3}$ as much as the second, and the second, only $\frac{1}{4}$ as much as the third. How much did each piece cost?

22. In an army consisting of 20,000 men, 3 times as many were wounded as were killed, and 4 times as many remained unhurt as were wounded. How many were killed, wounded, and unhurt respectively?

23. $\frac{1}{2}$ of A.'s money, $+ \frac{2}{3}$ of B.'s money equals $5500; and $\frac{2}{3}$ of B.'s money is 4 times $\frac{1}{2}$ of A.'s. How much money has each?

24. Herman and Byron together have 60 blocks, and Byron owns $\frac{2}{3}$ as many as Herman; how many has each?

25. Divide the number 60 into two parts, that shall be to each other as $\frac{1}{2}$ is to $\frac{3}{4}$?

26. Adelia and Louisa are to share 14 apples in the proportion of 4 to 3; how many ought each to receive?

27. The sum of Mary and Hezekiah's ages is 25 years; how old is each, provided Hezekiah is only $\frac{2}{3}$ as old as Mary?

28. Henry and his father can thrash out 35 bushels of oats in a day; how much does each, if Henry thrashes out only $\frac{2}{3}$ as much as his father?

29. A pole, whose length is 70 feet, is in the air and water; how much is in the air and water respectively, if $\frac{3}{4}$ of the length in the air equals the length in the water?

30. Divide the number 36 into two parts, that shall be to each other as 5 is to 4.

31. Divide the number 45 into two parts, that shall be to each other as 1 is to $\frac{4}{5}$.

32. A. and B. together own \$480; but A. owns only $\frac{3}{5}$ as much as B.; how much belongs to each?

33. A man died, and left \$7200 to be divided between his son and daughter, in the proportion of 1 to $\frac{2}{7}$. How much ought each to receive?

34. In a mixture of tea consisting of 48 pounds, there was $\frac{1}{3}$ as much poor, as good tea; how much of each kind was there?

35. A man bought a cow and a horse for \$96; the cow cost $\frac{3}{5}$ as much as the horse; how much was the cost of each?

36. Moses has only $\frac{3}{7}$ as many chestnuts as Aaron, and both have 40 quarts; how many quarts has each?

37. Divide the number 49 into two parts, that shall be to each other as 1 is to $\frac{2}{5}$.

38. A hound ran 60 rods before he caught a fox, and $\frac{2}{3}$ the distance the fox ran before he was caught, equaled the distance he was ahead when they started. How far did the fox run, and how far in advance of the hound was he, when the chase commenced?

39. The sum of two numbers is 140, and the larger is to the smaller as 1 is to $\frac{5}{9}$; what are the two numbers?

40. A. and B. together owe \$69, but B. owes only $\frac{11}{12}$ as much as A.; how much does each owe?

41. Thomas and Thornton found \$240, but could not agree about the division of it; they, therefore, threw it on the floor, and each got what he could; it so happened that Thomas got only $\frac{3}{5}$ as much as Thornton. How much did each get?

42. In a certain school consisting of 48 pupils, there are $1\frac{2}{7}$ times as many boys as girls; how many boys, and how many girls are in the school?

43. A gold and a silver watch were bought for \$160; the silver watch cost only $\frac{1}{7}$ as much as the gold one; how much was the cost of each?

44. Divide the number 17 into two parts, that shall be to each other as $\frac{2}{3}$ is to $\frac{3}{4}$.

45. A farmer had 180 sheep in two fields, and $\frac{1}{4}$ of the number in the first field equaled $\frac{1}{5}$ of the number in the second; how many in each field?

46. Divide 88 into two parts that shall be to each other as $\frac{2}{3}$ is to $\frac{1}{5}$.

47. $\frac{2}{5}$ of the distance a hare ran, after a hound started in pursuit, equaled the distance she was before the hound when they started; how far did the hare run before she was caught, provided the hound ran 80 rods to catch the hare?

48. A. and B. started from the same point, and ran in the same direction; B. ran 60 rods; then $\frac{1}{11}$ of the distance A. had run equaled the distance A. was ahead of B. How much did A. gain on B. in running 60 rods?

49. A fishing-rod, the length of which is 24 feet, is in two parts; $\frac{2}{3}$ of the longer part equals the length of the shorter. How long is each part?

50. A hound ran 90 rods before he caught a deer; the deer ran 44 times as far as it was ahead of the hound when they started, before it was caught. How far ahead of the hound was the deer when the chase commenced?

51. $\frac{2}{3}$ of A.'s number of sheep + $\frac{3}{4}$ of B.'s number, equals 900; how many sheep has each, provided $\frac{3}{4}$ of B.'s number is twice $\frac{2}{3}$ of A.'s number?

LESSON LIX.

1. A person had two silver cups, and only one cover for both. The first cup weighed 6 oz. If the first cup be covered, it will weigh twice as much as the second, but if the second cup be covered, it will

weigh 3 times as much as the first. What is the weight of the second cup and cover?

ANALYSIS.—By the last condition of the question, 3 times 6 oz., the weight of the first cup, or 18 ounces, equals the weight of the second cup and cover. Consequently, the two cups and cover weigh 18 + 6 ounces, which are 24 ounces. And by the first condition, the first cup and cover weigh twice as much as the second cup. Therefore, the 24 ounces must be divided into two parts, which are to each other as 2 to 1. One of these parts will be the weight of the second cup, and 2 the weight of the first cup and cover, &c.

2. A lady has two silver cups, and only one cover. The first cup weighs 8 ounces. The first cup and cover weigh 3 times as much as the second cup; and the second cup and cover 4 times as much as the first cup. What is the weight of the second cup and cover?

3. A man bought a hat, a coat, and a vest for $40. The hat cost $6; the hat and coat cost 9 times as much as the vest. What was the cost of each?

4. A boy bought a squirrel, a rabbit, and a bird. The squirrel cost 15 cts. The squirrel and rabbit cost twice as much as the bird; and the rabbit and bird cost 3 times as much as the squirrel. What was the cost of the bird and rabbit respectively?

5. A farmer bought a cow, an ox, and a horse; the cow cost $20. The cow and ox together cost 3 times as much as the horse; the ox and horse together cost 4 times as much as the cow. What was the cost of the ox and horse respectively?

6. A man bought two horses and a saddle. The younger horse cost $40. The saddle cost $\frac{1}{4}$ as much as both horses; and the younger horse cost $\frac{1}{3}$ as much as the other horse and saddle together. What did the saddle and older horse cost respectively?

7. A man traveled three successive days. The first day he traveled 30 miles, which was $\frac{1}{4}$ of the distance he traveled the other two days; and 4 times

the distance he traveled the second day equaled the distance he traveled the first and third days. How far did he travel each day?

8. A. is worth $1000, and B. and C. together are worth 9 times as much as A.; and C. is worth $\frac{1}{6}$ as much as A. and B. How much is B. and C. worth respectively?

9. A coat cost $20, and a vest and hat together cost 5 times as much as the coat; 3 times the cost of the vest equaled the cost of both coat and hat. What was the cost of the vest and the hat?

10. B.'s harness cost $120, which was $\frac{1}{3}$ of the cost of his horse and sleigh; and the harness and horse together cost twice as much as the sleigh. How much did the horse and sleigh cost respectively?

11. A pole in falling broke into three unequal pieces. The top piece was 8 feet long, which was $\frac{1}{5}$ of the length of the other two pieces; and 3 times the length of the bottom piece, equals the length of the other two pieces. How long was the pole, and how long was each piece?

12. Find the ages of A., B., and C., by knowing that A. is 20 years old, and that the sum of B. and C.'s ages is 4 times A.'s age; and that C.'s age is $\frac{1}{6}$ of the sum of A. and B.'s ages.

13. Find the fortunes of A., B., C., D., E., and F., by knowing that A. is worth $20, which is $\frac{1}{4}$ as much as B. and C. are worth, and that C. is worth $\frac{1}{5}$ as much as A. and B.; and, also, that if 19 times the sum of A., B., and C.'s fortunes were divided in the proportion of $\frac{3}{4}$, $\frac{1}{2}$, and $\frac{1}{3}$, it would respectively give $\frac{3}{4}$ of D.'s, $\frac{1}{2}$ of E.'s, and $\frac{1}{3}$ of F.'s fortune.

14. A. and B. dug 100 rods of ditch for $100. A. received 10 shillings a rod, and B. 6 shillings a rod. How many rods did each dig, provided each received $50. Ans. A. dug $37\frac{1}{2}$ rods, and B. $62\frac{1}{2}$.

Note.—$1 = 7s. 6d. in Pa. and N. J.; 6s. in New Eng. States, Va., Ky., and Tenn.; and 8s. in N. Y., Ohio, and N. C. currency.

LESSON LX.

REMARK.—In general business calculations of Interest, and in this book, 30 days to a month, and 12 of such months to a year, are reckoned, although such estimate would not be proper for a correct result in a question wherein should be stated any of the months which contain 31 days.

1. Reduce 2 years and 4 months to the fraction of a year.

ANALYSIS.—4 months is what part of a year? There are 12 months in 1 year; therefore, 1 month is $\frac{1}{12}$ of a year, and 4 months are 4 times $\frac{1}{12}$, which are $\frac{4}{12}$, or $\frac{1}{3}$ of a year. In $2\frac{1}{3}$ years, how many thirds? In one there are $\frac{3}{3}$; therefore, 3 times the number of whole years equal the number of thirds. 3 times 2 are 6, and $\frac{1}{3}$ added are $\frac{7}{3}$ of a year.

☞ REMARK.—*Always reduce a fraction to its lowest terms before performing any other operation with it.*

If the pupil can not readily discover the greatest number that will divide both *numerator* and *denominator*, without a remainder, he should continue to divide by any, or the least number that is contained in both numerator and denominator without a remainder, until the fraction is reduced to its lowest terms.

2. Reduce 1 year and 3 months to the fraction of a year.

3. Reduce 3 years and 5 months to the fraction of a year.

4. Reduce 4 years and 10 months to the fraction of a year.

5. Reduce 7 years and 9 months to the fraction of a year.

6. Reduce 8 years and 8 months to the fraction of a year.

7. Reduce 12 years and 7 months to the fraction of a year.

8. Reduce 11 years and 11 months to the fraction of a year.

9. Reduce 6 years and 6 months to the fraction of a year.

10. Reduce 9 years and 8 months to the fraction of a year.

11. Reduce 2 years 4 months and 15 days to the fraction of a year.

ANALYSIS.—15 days is what part of a month? There are 30 days in one month; therefore, 1 day is $\frac{1}{30}$ of a month, and 15 days are 15 times $\frac{1}{30}$, which are $\frac{15}{30}$, or $\frac{1}{2}$ of a month. $4\frac{1}{2}$ months, or $\frac{9}{2}$ months, is what part of a year? There are 12 months in 1 year; therefore, 1 month is $\frac{1}{12}$ of a year, and $\frac{1}{2}$ of a month is $\frac{1}{2}$ of $\frac{1}{12}$, which is $\frac{1}{24}$ of a year; and 9 halves are 9 times $\frac{1}{24}$, which are $\frac{9}{24}$, or $\frac{3}{8}$ of a year. $2\frac{3}{8}$ years, equal $\frac{19}{8}$ years.

12. Reduce 4 years 7 months and 6 days to the fraction of a year.

13. Reduce 5 years 9 months and 18 days to the fraction of a year.

14. Reduce 1 year 7 months and 18 days to the fraction of a year.

15. Reduce 2 years 7 months and 6 days to the fraction of a year.

REMARK.—Omitting the intermediate steps in the analysis, we have:

ANALYSIS.—6 days is $\frac{6}{30}$ or $\frac{1}{5}$ of a month. $7\frac{1}{5}$ months equal $\frac{36}{5}$ months. $\frac{36}{5}$ months equal $\frac{3}{5}$ of a year. $2\frac{3}{5}$ years equal $\frac{13}{5}$ years.

16. Reduce 3 years 3 months and 6 days to the fraction of a year.

17. Reduce 5 years 4 months and 24 days to the fraction of a year.

18. Reduce 6 years 5 months and 18 days to the fraction of a year.

19. Reduce 7 years 11 months and 6 days to the fraction of a year.

20. Reduce 10 years 10 months and 12 days to the fraction of a year.

LESSON LXI.

Interest is the money paid for the use of money, or its equivalent.

Per Cent., or *Rate per Cent.,* signifies by the hundred. Thus, by 6 per cent. is meant, 6 pounds on 100 pounds; $6 on $100, or $\frac{6}{100} = \frac{3}{50}$ of the quantity.

Per cent. is now generally written by business men thus, %. 7% meaning, 7 per cent.

The Principal is the sum on which interest is paid.

The Amount is the sum of the principal and interest.

1. At 4 per cent., what part of the principal equals the interest?

   ANALYSIS.—At 4 per cent., $\frac{4}{100}$, or $\frac{1}{25}$ of the principal, equals the interest.

2. At 2 per cent., what part of the principal equals the interest?

3. At 5 per cent., what part of the principal equals the interest?

4. At 6 per cent., what part of the principal equals the interest?

5. At 8 per cent., what part of the principal equals the interest?

6. At 10 per cent., what part of the cost equals the gain?

7. At 7 per cent., what part of the principal equals the interest?

8. At 12 per cent., what part of the cost equals the gain?

9. At 18 per cent., what part of the principal equals the interest?

10. What is the interest of $80 for 1 year, at 15 per cent.?

    ANALYSIS.—At 15 per cent., $\frac{15}{100}$, or $\frac{3}{20}$ of the principal, equals the interest. $\frac{3}{20}$ of $80 is $12, the interest.

11. What is the interest of $120 for 1 year, at 25 per cent.?

12. What is the interest of $510 for 1 year, at 20 per cent. ?

13. What is the interest of $750 for 1 year, at 24 per cent. ?

14. A man paid $120 for a wagon, and sold it at a gain of 30 per cent. ; how much was his gain ?

ANALYSIS.—If he gained 30 per cent., he gained $\frac{30}{100}$, or $\frac{3}{10}$ of the cost. $\frac{3}{10}$ of $120 is $36, the gain.

15. A tailor sold a coat that cost him $25, at a gain of 32 per cent. ; how much did he gain ?

16. A man sold a quantity of goods that cost him $340, at a gain of 75 per cent. ; how much did he gain ?

17. Edward spent 85 per cent. of $120 for a suit of clothes. How much did his clothes cost ?

18. Henry's watch cost $180, he sold it at a loss of 15 per cent. ; how much did he receive for it ?

19. A boy sold a quantity of candy that cost 50 cts., at a gain of 120 per cent. ; how much did he receive for them ?

20. Jacob sold a horse that cost him $240, at a loss of 25 per cent. ; how much did he receive for the horse ?

LESSON LXII.

REMARK.—The principal or cost is always 100 per cent.

1. If $\frac{3}{50}$ of the principal equals the interest, what is the rate per cent. ?

ANALYSIS 1ST.—If $\frac{3}{50}$ of the principal equals the interest, the rate per cent. is $\frac{3}{50}$ of 100 per cent., which is 6 per cent.

ANALYSIS 2D.—If the interest of 1 cent is $\frac{3}{50}$ of a cent, the interest of 100 cents is 100 times $\frac{3}{50}$, or $\frac{300}{50}$, or 6 cents.

2. If $\frac{1}{50}$ of the principal equals the interest, what is the rate per cent. ?

3. If $\frac{2}{25}$ of the principal equals the interest, what is the rate per cent. ?

4. If $\frac{9}{50}$ of the cost equals the gain, what is the rate per cent.?

5. If $\frac{3}{25}$ of the cost equals the gain, what is the rate per cent.?

6. If $\frac{1}{25}$ of the principal equals the interest, what is the rate per cent.?

7. If $\frac{1}{15}$ of the cost equals the gain, what is the rate per cent.?

8. If the interest of $44 for 1 year is $4, what is the rate per cent.?

ANALYSIS.—If the interest of $44 is $4, $\frac{4}{44}$, or $\frac{1}{11}$ of the principal equals the interest. Therefore, the rate per cent. is $\frac{1}{11}$ of 100 per cent., which is $9\frac{1}{11}$ per cent.

9. If the interest of $72 for 1 year is $6, what is the rate per cent.?

10. If the interest of $96 for 1 year is $12, what is the rate per cent.?

11. B. bought a horse for $100, and sold it for $109; how much did he gain per cent.?

12. A woman bought a quantity of oranges for 75 cents, and sold them for 84 cents; how much did she gain per cent.?

13. A merchant bought a quantity of books for $200, and sold them for $228; how much did he gain per cent.?

14. If by laying out $37, I gain a sum equal to $\frac{2}{5}$ of it, what do I gain per cent.?

15. Harvey bought a hogshead of molasses for $25, and sold it for 31\frac{1}{2}$; how much did he gain per cent.?

16. Bought a knife for 37 cents, and sold it for 57$\frac{1}{2}$ cents; what was the gain per cent.?

17. A stationer sold a quantity of paper for $\frac{5}{4}$ of what it cost; how much did he gain per cent.?

18. James received for his horse $\frac{7}{6}$ of what it cost; how much did he gain per cent.?

7

19. A man sold a barrel of pork for $\frac{7}{5}\frac{1}{0}$ of what it cost; how much did he gain per cent.?

20. The interest of $500 for 4 years is $240; what is the rate per cent.?

ANALYSIS.—If the interest of $500 for 4 years is $240, for 1 year it is $\frac{1}{4}$ of $240, or $60. Therefore, $\frac{60}{500}$, or $\frac{3}{25}$ of the principal, equals the interest. Hence, the rate per cent. is $\frac{3}{25}$ of 100%, which is 12%.

21. A man being asked at what per cent. his money was on interest, replied, I receive $120 interest in 10 years for $240; what was his rate per cent.?

22. A. bought a horse for $150, and sold it for $180? what was his gain per cent.?

23. Elisha bought 10 horses for $800, and sold 8 of them for what all cost; what was his gain per cent.?

24. $\frac{7}{50}$ of the money which C. paid for books, is $\frac{1}{8}$ of what he gained by selling them. How much did he gain per cent.?

25. $\frac{8}{25}$ of the money that I have on interest, is 4 times the yearly interest received. What is the rate per cent.?

26. $1\frac{3}{45}$ of the cost of A.'s merchandise, is $\frac{4}{5}$ of what he gained when he sold it. What was his gain per cent.?

27. $\frac{4}{75}$ of the cost of B.'s wagon was $\frac{2}{3}$ of what he gained by selling it; what did he gain per cent.?

28. A book was sold for $\frac{2}{3}$ of $\frac{6}{8}$ of what it cost; what was the loss per cent.?

29. $\frac{3}{4}$ of $\frac{8}{5}$ of the cost of a sleigh, was what the sleigh sold for; what was the gain per cent.?

30. A merchant bought a quantity of goods for $860, and sold them for $1075; how much did he gain per cent.?

LESSON LXIII.

NOTE.—The expression of a certain rate per cent. (%) of interest means per annum, *i. e.*, for a year, unless otherwise specified.

1. At 5 per cent. for 4 years, what part of the principal equals the interest?

  ANALYSIS.—If the interest of $1 for 1 year is 5 cents, for 4 years it is 4 times 5 cents, or 20 cents. Therefore, $\frac{20}{100}$, or $\frac{1}{5}$ of the principal, equals the interest.

2. At 6 per cent. for 5 years, what part of the principal equals the interest?

3. At 3% for 2 years, what part of the principal equals the interest?

4. At 4% for 3 years, what part of the principal equals the interest?

5. At 6% for 3 years, what part of the principal equals the interest?

6. At 4% for 3 years, what part of the principal equals the interest?

7. At 9% for 6 years, what part of the principal equals the interest?

8. At 8% for 5 years, what part of the principal equals the interest?

9. At 4% for 6 years, what part of the principal equals the interest?

10. At 6% for 8 years, what part of the principal equals the interest?

11. At 10% for 5 years, what part of the principal equals the interest?

12. At 6% for 4 years and 8 months, what part of the principal equals the interest?

REMARK.—It is expected that pupils understand the lessons before this, and are, therefore, prepared to arrive at results, without giving the analysis of all parts of the question.

ANALYSIS.—8 months are $\frac{8}{12}$, or $\frac{2}{3}$ of a year. $4\frac{2}{3}$ years equal $\frac{14}{3}$ years. If the interest of $1 for 1 year is 6 cents, for $\frac{14}{3}$ years it is $\frac{14}{3}$ times 6 cents, or 28 cents. Therefore, $\frac{28}{100}$, or $\frac{7}{25}$ of the principal, equals the interest.

13. At 4 per cent. for 6 years and 6 months, what part of the principal equals the interest?

14. At 6 per cent. for 5 years and 4 months, what part of the principal equals the interest?

15. At 10½ per cent. for 1 year and 6 months, what part of the principal equals the interest?

16. At 4⅔ per cent. for 9 years, what part of the principal equals the interest?

17. At 3⅗ per cent. for 2 years and 2 months, what part of the principal equals the interest?

18. At 6¼ per cent. for 4 months and 24 days, what part of the principal equals the interest?

19. At 7⅕ per cent. for 10 months, what part of the principal equals the interest?

20. At $3\frac{7}{19}$% for 2 years 4 months and 15 days, what part of the principal equals the interest?

LESSON LXIV.

1. What is the interest of $50 for 4 years, at 6%?

ANALYSIS 1ST.—The interest of $1 for 4 years, at 6%, is 24 cents; and for $50 it is 50 times 24 cents, or $12.

ANALYSIS 2D.—The interest of $1 for 4 years, at 6%, is 24 cents. Therefore, $\frac{24}{100}$, or $\frac{6}{25}$ of the principal, equals the interest. $\frac{6}{25}$ of $50 is $12, the interest.

2. What is the interest of $10 for 2 years, at 5%?
3. What is the interest of $48 for 6 years, at 5%?
4. What is the interest of $70 for 7 years, at 5%?
5. What is the interest of $68 for 5 years, at 6%?
6. What is the interest of $70 for 2 years, at 5%?
7. What is the interest of $75 for 5 years, at 3%?
8. What is the interest of $120 for 8 years, at 5 per cent.?

9. What is the interest of $100 for 10 years, at 6 per cent.?

10. What is the interest of $140 for 12 years, at 5 per cent.?

11. What is the interest of $150 for 5 years, at 3 per cent.? ~

12. What is the interest of $145 for 6 years, at 5 per cent.?

13. What is the interest of $200 for 10 years, at 8 per cent.?

14. What is the interest of $250 for 3 years, at 8 per cent.? .

15. What is the interest of $220 for 11 years, at 10 per cent.?

16. What is the interest of $500 for 9 years, at 8 per cent.?

17. What is the interest of $250 for 12 years, at 6 per cent.?

18. What is the interest of $500 for 8 years, at 12 per cent.?

19. What is the interest of $200 for 9 years, at 3 per cent.?

20. What is the interest of $405 for 10 years, at 8 per cent.?

21. What is the interest of $50 for 2 years and 2 months, at 6 per cent.?

ANALYSIS.—2 months is $\frac{1}{6}$ of a year. $2\frac{1}{6}$ years equals $\frac{13}{6}$ years. If the interest of $1 for 1 year is 6 cents, for $\frac{13}{6}$ years it is $\frac{13}{6}$ times 6 cents, or 13 cents. Therefore, $\frac{13}{100}$ of the principal equals the interest. $\frac{13}{100}$ of $50 is $\$\frac{13}{2}$, or $6.50, the interest.

22. What is the interest of $25 for 4 years and 3 months, at 4 per cent.?

23. What is the interest of $80 for 5 years and 5 months, at 6 per cent.?

24. What is the interest of $60 for 8 years and 6 months, at 6 per cent.?

25. What is the interest of $240 for 3 years and 9 months, at 6 per cent.?

26. What is the interest of $75 for 4 years and 8 months, at 9 per cent.?

27. What is the interest of $50 for 2 years and 9 months, at 6 per cent.?

28. What is the interest of $80 for 12 years and 10 months, at 6 per cent.?

29. What is the interest of $69 for 8 years and 4 months, at 2 per cent.?

30. What is the interest of $60 for 4 years and 8 months, at 3 per cent.?

31. What is the interest of $600 for 2 years 4 months and 15 days, at 4 per cent.?

> ANALYSIS.—15 days is $\frac{1}{2}$ of a month. $4\frac{1}{2}$ months equals $\frac{9}{2}$ months. $\frac{9}{2}$ months equals $\frac{9}{24}$, or $\frac{3}{8}$ of a year. $2\frac{3}{8}$ years equal $\frac{19}{8}$ years. If the interest of $1 for 1 year is 4 cents, for $\frac{19}{8}$ years it is $\frac{19}{8}$ times 4 cents, or $\frac{19}{2}$ cents. Therefore, $\frac{19}{200}$ of the principal equals the interest. $\frac{19}{200}$ of $600 is $57, the interest required.

32. What is the interest of $300 for 5 years 9 months and 18 days, at 5 per cent.?

33. What is the interest of $550 for 4 years 7 months and 6 days, at 10 per cent.?

34. What is the interest of $500 for 1 year 7 months and 18 days, at 6 per cent.?

35. What is the interest of $250 for 3 years 7 months and 6 days, as 4 per cent.?

36. What is the interest of $250 for 3 years 3 months and 6 days, at 6 per cent.?

37. What is the interest of $50 for 6 years 4 months and 24 days, at 5 per cent.?

38. What is the interest of $75 for 2 years 11 months and 6 days, at 15 per cent.?

39. What is the interest of $150 for 2 years 6 months and 12 days, at 15 per cent.?

40. What is the interest of $300 for 2 years 9 months and 18 days, at $1\frac{1}{2}$ per cent.?

LESSON LXV.

1. What is the *amount* of $75 for 2 years, at 6 per cent. per annum ?

> ANALYSIS.— If the interest of $1 for 1 year is 6 cents, for 2 years it is 2 times 6 cents, or 12 cents. Therefore, $\frac{12}{100}$, or $\frac{3}{25}$, of the principal equals the interest. $\frac{3}{25}$ of $75 is $9, the interest ; to which add $75, the principal, and we have $84, the amount.

2. What is the amount of $90 for 3 years, at 7 per cent. ?

3. What is the amount of $100 for 4 years, at 5 per cent. ?

4. What is the amount of $160 for 10 years, at 5 per cent. ?

5. What is the amount of $160 for 8 years, at 5 per cent. ?

6. What is the amount of $200 for 12 years, at 5 per cent. ?

7. What is the amount of $210 for 2 years and 6 months, at 4 per cent. ?

8. What is the amount of $250 for 4 years and 3 months, at 8 per cent. ?

9. What is the amount of $240 for 4 years and 2 months, at 3 per cent. ?

10. What is the amount of $500 for 3 years 3 months and 6 days, at 6 per cent. ?

11. What is the amount of $200 for 5 years 4 months and 24 days, at 5 per cent. ?

LESSON LXVI.

1. What principal will in 4 years, at 6 per cent., give $12 interest ?

> ANALYSIS.— If the interest of $1 for 1 year is 6 cents, for 4 years it is 4 times 6 cents, or 24 cents. Therefore, $\frac{24}{100}$, or $\frac{6}{25}$, of the principal equals the interest, which is $12. If $\frac{6}{25}$ of the principal is $12, $\frac{1}{25}$ of the principal is $\frac{1}{6}$ of $12, or $2 ; and $\frac{25}{25}$, or the principal, is 25 times $2, or $50.

2. What principal will in 6 years, at 4 per cent., give $36 interest?

3. What principal will in 4 years, at 5 per cent., give $30 interest?

4. What principal will in 8 years, at 7 per cent., give $42 interest?

5. What principal will in 10 years, at 7 per cent., give $140 interest?

6. What principal will in 4 years and 6 months, at 6 per cent., give $54 interest?

7. What principal will in 4 years and 3 months, at 5%, give $102 interest?

8. What principal will in 4 years and 3 months, at 8%, give $51 interest?

9. How much money has that man on interest, who, at the expiration of 4 years and 4 months, at 6 per cent., receives $260 interest?

10. At the expiration of 2 years and 4 months, at 6 per cent., a man received $49 interest. How much money had he on interest?

11. A. is worth twice as much as B., and the interest of their united fortunes for 4 years and 2 months, at 6%, is $600. How much is each worth?

12. The interest on the cost of B.'s store and house, for 1 year and 6 months, at 4%, would be $270. What was the cost of each, provided the store cost ¼ as much as the house?

13. If the money B. paid for a sheep, a cow, and a horse, was put on interest for 4 years and 6 months, at 4 per cent., it would give $18 interest. What was the cost of all, and of each, provided the sheep cost ⅓ as much as the cow, and the cow, ½ as much as the horse?

NOTE.—By the condition of the question, the cow must have cost 3 times as much as the sheep, and the horse 2 times as much as the cow, or 6 times as much as the sheep.

LESSON LXVII.

1. What principal will in 4 years, at 5 per cent., amount to $360 ?

> ANALYSIS.—If the interest of $1 for 1 year is 5 cents, for 4 years it is 4 times 5 cents, or 20 cents. Therefore, $\frac{20}{100}$, or $\frac{1}{5}$, of the principal equals the interest; to which add $\frac{5}{5}$, the principal, and we have $\frac{6}{5}$ of the principal equal to the amount, $360. If $\frac{6}{5}$ of the principal is $360, $\frac{1}{5}$ of the principal is $\frac{1}{6}$ of $360, which is $60, and $\frac{5}{5}$ (the principal), is 5 times $60, which is $300.

2. What principal will in 3 years, at 6 per cent., amount to $118 ?

3. What principal will in 6 years, at 10 per cent., amount to $120 ?

4. What principal will in 10 years, at 7 per cent., amount to $170 ?

5. What principal will in 4 years, at 5 per cent., amount to $660 ?

6. A. is worth $\frac{1}{3}$ as much as B.; and the interest on their united fortunes for 2 years, at 5 per cent., is $880. What is the fortune of each ?

7. A merchant sold a quantity of cloth for $214, and thereby gained 7 per cent.; what did the cloth cost him ?

8. What principal will in 2 years, at 7 per cent., amount to $1140 ?

9. What principal will in 10 years and 8 months, at 9 per cent., amount to $490 ?

10. The amount due on a note, which had been on interest 6 years and 2 months, at 6%, was $274; what was the face of the note ?

11. What principal will in 12 years and 9 months, at 4%, amount to $302 ?

12. If $\frac{1}{2}$ of A.'s fortune for 4 years and 6 months, at 6%, amounts to $127, what is his whole fortune ?

13. If ⅔ of B.'s fortune, being put on interest for 3 years 3 months and 6 days, at 15 per cent., amounted to $149, what was his whole fortune?

14. Mary, being asked how much money she had on interest, and at what per cent., replied; the *principal* and *rate %* are such that in 5 years the *amount* would be $750, and in 7 years, $810. What was the principal and the rate per cent.?

15. A man sold two horses for $240, losing on the first 20 per cent., gaining on the other 20%; what was the value of each horse, provided he received for the second 3 times as much as for the first?

16. The amount of Robert's capital for a certain time, at 4%, was $360, and for the same time, at 7%, it was $405; required his principal and the time.

LESSON LXVIII.

1. In what time will $40, at 6 per cent., give $12 interest?

ANALYSIS.—If the interest of $40 is $12, $1\frac{2}{6}$, or $\frac{3}{10}$, of the principal equals the interest. If the interest of $1 for 1 year is $\frac{3}{10}$ of a dollar, of $100 it is 100 times $\frac{3}{10}$, or $30. If it require 1 year for $100 to give $6 interest, to give $30 interest it will require as many years as $6 is contained times in $30, or 5 years.

2. In what time will $60, at 5 per cent., give $18 interest?

3. In what time will $90, at 7 per cent., give $27 interest?

4. In what time will $100, at 6 per cent., give $10 interest?

5. In what time will $120, at 10 per cent., give $120 interest?

6. In what time will $250, at 6 per cent., give $20 interest?

7. In what time will $40, at 7%, give $8.40 interest?

8. In what time, at 8%, will $30 give $9.60 interest?

9. In what time, at 6%, will $10 give $2.40 interest?

10. In what time, at 4%, will $20 give $5.60 interest?

LESSON LXIX.

1. At what rate per cent., will $50, in 1 year and 6 months (or $1\frac{1}{2}$ years), give $6 interest?

ANALYSIS.—If the interest of $50 for $1\frac{1}{2}$, or $\frac{3}{2}$ years, is $6, for $\frac{1}{2}$ of a year it is $\frac{1}{3}$ of $6, or $2; and for $\frac{2}{2}$, or 1 year, it is 2 times $2, or $4. Therefore, $\frac{4}{50}$, or $\frac{2}{25}$, of the principal equals the annual interest. Hence, the rate per cent. is $\frac{2}{25}$ of 100% = 8%.

2. At what rate per cent., will $40 annually give $2 interest?

3. At what rate per cent., will $80 annually give $3.20 interest?

4. At what rate per cent., will $120 annually give $12 interest?

5. At what rate per cent., will $120 in 4 years, give $20 interest?

6. At what rate per cent., will $100 in 3 years, give $30 interest?

7. At what rate per cent., will $5 in 14 years, give $7 interest?

8. At what rate per cent., will $25 in 1 year and 9 months, give $3.50 interest?

9. At what rate per cent., will $80 in 5 years and 8 months, give $34 interest?

10. At what rate per cent., will $500 in 7 years and 6 months, give $15 interest?

11. At what rate per cent., will $600 in 2 years 4 months and 15 days, give $57 interest?

LESSON LXX.

1. At what rate per cent., will $10 in 4 years, amount to $12 ?

REMARK.—From the amount subtract the principal, and the remainder will be the interest. Then proceed as in the preceding lesson.

2. At what rate per cent., will $12 in 3 years, amount to $13.44 ?

3. At what rate per cent., will $20 in 6 years, amount to $26 ?

4. At what rate per cent., will $24 in 10 years, amount to $36 ?

5. At what rate per cent., will $30 in 7 years, amount to $36.30 ?

6. At what rate per cent., will $50 in 10 years, amount to $75 ?

7. At what rate per cent., will $36 in 5 years, amount to $39.60 ?

8. At what rate per cent., will the interest for 20 years equal a given principal ?

ANALYSIS.—With the interest at 100%, it would equal a given principal, or a given principal will double itself in 1 year, at 100 per cent.; and in 20 years, at $\frac{1}{20}$ of 100 per cent., or 5 per cent.

9. At what rate per cent., will a given principal double itself, in 4 years ?

10. At what rate per cent., will a given principal double itself, in 3 years ?

11. At what rate per cent., will the interest equal a given principal in 5 years ?

12. At what rate per cent., will $80 in 7 years give $80 interest ?

13. At what rate per cent., will $640 in 6 years give $640 interest ?

14. At what rate per cent., will 25 cents in 8 years give 25 cents interest?

15. At what rate per cent., will $97 in 9 years give $97 interest?

16. At what rate per cent., will $372 in 25 years give $372 interest?

17. At what rate per cent., will $1 in 30 years give $1 interest?

18. At what rate per cent., will $15 in 12½ years give $15 interest?

19. At what rate per cent., will $42 in 14⅞ years give $42 interest?

20. At what rate per cent., will 5 cents in 16⅝ years give 5 cents interest?

LESSON LXXI.

1. In what time will a given principal double itself, at 5 per cent.?

ANALYSIS.—A given principal will double itself in 100 years, at 1 per cent., and at 5 per cent., in ⅕ of 100 years, which is 20 years.

2. In what time will a given principal double itself, at 4 per cent.?

3. In what time will $25, at 3 per cent., give $25 interest?

4. In what time will $275, at 6 per cent., give $275 interest?

5. In what time will the interest equal a given principal, at 2 per cent.?

6. In what time will $4, at 7%, give $4 interest?

7. In what time will $94, at 9%, give $94 interest?

8. In what time will 5 cents, at 8 per cent., give 5 cents interest?

9. In what time will $3⅝, at 10 per cent., give $3⅝ interest?

10. In what time will 1 dime, at 12½ per cent., give 1 dime interest?

LESSON LXXII.

1. Bought a bushel of grass-seed for $5, and sold it for $7; what was the gain per cent.?

   ANALYSIS.—Since it was bought for $5, and sold for $7, the gain was $7—$5, which is $2. Therefore, ⅖ of the cost equals the gain. Hence, the gain per cent. was ⅖ of 100 per cent., which is 40 per cent.

2. A book was bought for $2, and sold for $3; what was the gain per cent.?

3. A shawl cost $5, and was sold for $8; what was the gain per cent.?

4. A cow was bought for $20, and sold for $25; what was the gain per cent.?

5. A merchant bought a hogshead of molasses for $80, and sold it for $95; what did he gain per cent.?

6. A barrel of pork cost $12, and was sold for $11; what was the loss per cent.?

7. A horse was bought for $140, and sold for $60; what was the loss per cent.?

8. Bought an orange for 4 cents, and sold it for 6 cents; what was the gain per cent.?

9. Bought a melon for 15 cents, and sold it for 20 cents; what was the gain per cent.?

10. Bought a book for 5 dimes, and sold it for 8 dimes; what was the gain per cent.?

11. Bought a hectometre of silk for $120, and sold it for $200; what was the gain per cent.?

12. A boy sold melons, at the rate of 10 cents each

$\frac{1}{5}$ of which equaled his gain; how much would he have gained per cent., if he had sold them at 12 cents each?

13. A merchant sold sugar for $80 a hogshead, and thereby cleared $\frac{1}{10}$ of this money; if he had sold it at $92 a hogshead, what would he have gained per cent.?
14. A quantity of cloth was bought for $36, and sold for $43; what was the gain per cent.?
15. A horse was bought for $100, and sold for $95; what was the loss per cent.?
16. A man re-sold a kilogramme of wheat for $120, and cleared $\frac{1}{5}$ of its cost; how much money would he have lost per cent., if he had sold it for $80?
17. What per cent. of $\frac{1}{3}$ is $\frac{1}{6}$? Of $\frac{2}{3}$ is $\frac{1}{6}$? Of $\frac{2}{5}$ is $\frac{1}{20}$? Of $\frac{3}{4}$ is $\frac{2}{3}$? Of $\frac{7}{8}$ is $\frac{2}{3}$? Of $2\frac{1}{2}$ is $\frac{2}{3}$? Of $3\frac{1}{5}$ is $2\frac{1}{10}$?
18. $\frac{2}{3}$ of $6 is what per cent. of $\frac{4}{5}$ of $100?
19. $\frac{5}{7}$ of $28 is $\frac{2}{3}$ of what per cent. of $\frac{5}{6}$ of $300?
20. Walter sold a horse for $120, and thereby gained $\frac{1}{5}$ of its cost; what would he have lost per cent. by selling it for $80?

LESSON LXXIII.

1. A man bought a cow for $20; for what must he sell her, to gain 5 per cent.?

ANALYSIS.—If he gain 5 per cent., he gains $\frac{5}{100}$, or $\frac{1}{20}$ of the cost. $\frac{1}{20}$ of $20 is $1, the gain. Therefore, to gain 5 per cent., he must sell the cow for $20 + $1, or $21.

2. A man bought a yoke of oxen for $100; how must he sell them, to gain 6 per cent.?
3. A man bought a barrel of flour for $10; for what must he sell it, to gain 10 per cent.?

4. A gallon of wine was bought for 20 dimes; how must it be sold a pint, to gain 20 per cent.?

5. A hogshead of molasses cost $20; for what ought it to be sold a gallon, to gain 40 per cent.?

6. B. bought a horse for $80, and by selling it, lost 5 per cent.; for what did he sell it?

7. A wagon cost $140, and was sold at a loss of 5 per cent.; for what was it sold?

8. A merchant, by selling 40 metres of cloth for $164, lost 20 per cent.; what did it cost per metre?

9. If 3 decalitres and 5 litres of wine cost $140, how must it sell per litre to gain 20%?

10. B. lost 5 per cent. by selling a hectolitre of rum, which cost $80; for what did he sell it a litre?

LESSON LXXIV.

1. What principal will, in 4 years, at 5 per cent., amount to $60?

ANALYSIS.—If the interest of $1 for 1 year is 5 cents, for 4 years it is 4 times 5 cents, or 20 cents. Therefore, $\frac{20}{100}$, or $\frac{1}{5}$ of the principal equals the interest; to which add $\frac{5}{5}$, the principal, and we have $\frac{6}{5}$ of the principal equal to the amount, or $60. If $\frac{6}{5}$ of the principal is $60, $\frac{1}{5}$ of the principal is $\frac{1}{6}$ of $60, which is $10; and $\frac{5}{5}$ (the principal), is 5 times $10, which are $50.

2. What principal will, in 3 years, at 6 per cent., amount to $118?

3. What principal will, in 5 years, at 6 per cent., amount to $130?

4. What principal will, in 7 years, at 5 per cent., amount to $81?

5. What principal will, in 9 years, at 8 per cent., amount to $86?

6. What principal will, in 3¾ years, at 8 per cent., amount to $260?

7. What principal will, in 4⅔ years, at 6 per cent., amount to $640 ?

8. What principal will, in 5⅚ years, at 7 per cent., amount to $42 ?

9. What principal will, in 6⅔ years, at 7 per cent., amount to $87 ?

10. What principal will, in 8⅜ years, at 6 per cent., amount to $76 ?

The present worth of a debt payable at some future time, without interest, is such a sum as will, in the given time, and at the given rate per cent., amount to the debt. Hence, the *present worth* of any sum of money, payable at some future time without interest, may be found in the same way that we found the *principal,* when the amount, time, and rate per cent. were given.

See the Analysis of this Lesson.

11. What is the present worth of $26, due 5 years hence, at 6 per cent. ? *Ans.* $20.

12. What is the present worth of $14, due 8 years hence, at 5 per cent. ?

13. What is the present worth of $110, due 5 years hence, at 5 per cent. ?

14. What is the present worth of $86, due 8 years hence, at 9 per cent. ?

15. What is the present worth of $102, due 9 years hence, at 4 per cent. ?

16. What is the present worth of $72, due 4 years hence, at 5 per cent. ?

Discount equals the amount minus the present worth.

17. What is the discount on $46, due 3 years hence, at 5 per cent. ?

18. What is the discount on $54, due 5 years hence, at 7 per cent. ?

19. What is the discount on $65, due 5 years hence, at 6 per cent. ?

20. What is the discount on $93, due 3 years hence, at 8 per cent. ?

21. What is the present worth of $186, due $4\frac{4}{5}$ years hence, at 5 per cent.?
22. What is the present worth of $66, due $5\frac{1}{3}$ years hence, at 6 per cent.?
23. What is the present worth of $128, due $4\frac{2}{3}$ years hence, at 6 per cent.?

LESSON LXXV.

1. If I sell cloth at $2.50 a yard, and thereby gain 25 per cent., what did it cost a yard?

   ANALYSIS.—If I gain 25 per cent., I gain $\frac{1}{4}$ of the cost; to which add $\frac{4}{4}$, the cost, and I have $\frac{5}{4}$ of the cost equal to $2.50. If $\frac{5}{4}$ of the cost is $2.50, $\frac{1}{4}$ of the cost is $\frac{1}{5}$ of $2.50, or 50 cents; and $\frac{4}{4}$ (the cost) is 4 times 50 cents, which are 200 cents, or $2.

2. A horse was sold for $38, which was at a loss of 5%; what did the horse cost?
3. If I sell cloth at $2.50 a yard, and thereby gain 25%, how must I sell it a yard to lose 20%?
4. If I sell cloth at $4.40 a metre, and thereby gain 10%, how ought I sell it to lose 25%?
5. If by selling a piece of cloth for $46, I gain 15%, how ought I to have sold it, to have lost 30%?
6. A. sold his horse for $105, and thereby gained 5%; for what ought he to have sold it, to have lost 10%?
7. A farm was sold for $495, which was 10% less than what it was worth; for what ought it to have been sold, to have received 40% more than its value?
8. A mechanic lost 20% on the cost of a wagon, by selling it for $40; for what ought it to have been sold, to have gained 30 per cent.?
9. A hectare was sold for $90, which was 10% less than its value; what would have been the gain per cent., if it had been sold for $120?

10. A farm was sold for $690, which was 8% less than its value; what would have been the gain % on its value, if it had been sold for $850 ?

11. A book was sold for 90 cents, which was 10% less than its value; what would have been the gain % on its value, if it had been sold for $1.50?

12. A man sold two watches, at $12 each; on one he gained 50%, and on the other he lost 50%. Did he gain or lose by the bargain, and how much ?

13. An individual sold two gold pencils, at $6 each; on one he gained 20%, and on the other he lost 20%. Did he gain or lose, and how much ?

14. A farmer sold two horses at $210 each; for one he received 25% more than its value, and for the other 25% less than its value. Did he gain or lose by the sale of both, and how much ?

15. A merchant sold a quantity of cloth for $280, and by so doing lost 60%; he then sold another quantity for $80, and gained 60%. Did he gain or lose by the entire sale, and how much ?

LESSON LXXVI.

1. An individual was ordered to collect $190, and his own fee, which was to be 5 per cent. on all the money collected. How much should he receive?

ANALYSIS.—He is to receive 5 per cent., or $\frac{5}{100}$, or $\frac{1}{20}$ of all he collects. $\frac{20}{20}$, all he collects, minus $\frac{1}{20}$, his fee, equals $\frac{19}{20}$ of all he collects, or $190, the amount he is to pay his employer. If $\frac{19}{20}$ of what he collects equals $190, $\frac{1}{20}$ is $\frac{1}{19}$ of $190, which is $10; and $\frac{20}{20}$ (what he collects) is 20 times $10, or $200. Therefore he must receive $200—$190=$10.

2. How much ought A. to receive for collecting $90 and his own fee of 10 per cent. on all he collects ?

3. What amount of money will be sufficient to pay a debt of $38 and the collector's fee, which is 5 per cent. on all the money collected?

4. How much cider must that man make to bring away 15 barrels, after the owner of the mill receives 16⅔ per cent. of all he has made?

5. How much grain must a farmer take to mill, that he may bring away the flour of 1 bushel, after the miller has taken 10 per cent. of all he took there?

MISCELLANEOUS QUESTIONS.

1. At 5 per cent. for 4 years, what part of the principal equals the interest?

2. In how many years, at 4 per cent., will the interest on a given principal amount to the same sum, as it will in 8 years, at 6 per cent.?

3. At what rate per cent. will the interest on a given principal, in 14 years, amount to the same sum, as it will in 12 years at 7 per cent.?

4. If $\frac{1}{20}$ of the principal equals the interest, what is the rate per cent.?

5. The rent of B.'s farm, for 8 years, equaled $\frac{32}{30}$ of its value; what per cent. did he annually receive on the value of his farm?

6. What is the interest of $75, for 5⅜ years, at 6 per cent.?

7. What principal will, in 7½ years, at 8 per cent., give $24 interest?

8. What principal will, in 4⅖ years, at 5 per cent., amount to $155?

9. At what rate per cent. will a given principal double itself in 12½ years?

10. The interest of A.'s and B.'s fortune, for 8 years, at 5 per cent., is $420; what is the fortune of each, provided A.'s fortune is twice B.'s?

11. The interest of $\frac{2}{3}$ of A.'s and $\frac{3}{4}$ of B.'s fortune, for 7 years, at 5 per cent., is $2100; what is each of their fortunes, provided $\frac{2}{3}$ of A.'s fortune equals $\frac{3}{4}$ of B.'s?

12. B. sold his horse for $\frac{1}{2}$ of $1\frac{1}{2}$ times what it cost; what did he lose per cent.?

13. What is the interest of $540 for 4 years, at 5 per cent.?

14. What is the interest of $180, for 5 years and 9 months, at $6\frac{2}{3}$ per cent.?

15. What principal will in 4 years 7 months and 6 days, at $6\frac{1}{4}$ per cent., amount to $412?

16. The interest of the cost of B.'s horse, sleigh, and wagon, for 6 years, at 5 per cent., is $69. What is the cost of each, provided their prices are to each other as $\frac{1}{2}$, $\frac{2}{3}$, and $\frac{3}{4}$?

17. What principal will, in 8 years and 8 months, at $7\frac{2}{5}$ per cent., amount to $419?

18. What principal will, in 5 years 9 months and 18 days, at 10 per cent., give $116 interest?

19. In what time will $420, at 5 per cent., give $147 interest?

20. If the interest of $200, for 1 year and 6 months, is $18, what is the rate per cent.?

21. At what rate per cent., will $500, in 4 years and 9 months, give $190 interest?

22. At what rate per cent., will $500, in 22 years and 6 days, amount to $1821?

23. At what rate per cent. will a given principal double itself, in 20 years?

24. In what time will the amount be double a given principal, at $12\frac{1}{2}$ per cent.?

25. At what rate per cent. will the amount be double a given principal, in 6 years and 8 months?

26. A horse was bought for $60, and sold for $90; what was the gain per cent.?

27. A basket containing 39 oranges, cost $1.20; how must they be sold each to gain 30 per cent.?

28. If 1 litre of wine cost 40 cents, how must it be sold a decilitre to gain 20 per cent.?

29. What is the present worth of $68, due 10 years hence, at 7 per cent.?

30. What is the discount on $162, due 10 years and 4 months hence, at 6 per cent.?

31. What is the present worth of $87, due 3½ years hence, at 5 per cent.?

32. If a hogshead of molasses containing 84 gallons cost $30, how must it be sold a gallon, to gain 40 per cent.?

33. The money I have on interest, in 9 years, at 10 per cent., amounts to $190; what is the principal?

34. When money was worth 6 per cent., I bought $400 worth of goods; 6 months afterwards I sold them, and gained 10 per cent. on the cost. How much did I gain?   *Ans.* $28.

35. A speculator bought a horse for $36, and sold it for 25 per cent. more than he gave for it; which, however, was 10 per cent. less than what he asked for it. How much did he ask for the horse?

36. A gentleman being asked how much money he had on interest, replied, that if instead of 6 per cent. he should receive 10 per cent., he would receive $268 interest more than he then did. How much money had he on interest?

37. A merchant bought broadcloth for $1.20 a yard, and sold it for 33⅓ per cent. more than he gave for it; which, however, was 33⅓ per cent. less than his marked price for it. How much was his marked price per yard?

38. A Frenchman sold a hectometre of cloth for $120, and gained 50 per cent. He sold another hectometre for $120, and lost 50 per cent. Did he gain or lose by both sales? how much?

39. B. sold a horse for \$60, and gained 20 per cent. He then sold another horse for \$60, and lost 60 per cent. Did he gain or lose, and how much?

40. The interest on $1\frac{1}{3}$ times A.'s, and $\frac{3}{5}$ of B.'s fortune, for 8 years, at 5 per cent., is \$520. What is the fortune of each, provided $1\frac{1}{3}$ times A.'s fortune equals $\frac{3}{5}$ of B.'s?

41. $\frac{2}{3}$ of D.'s fortune added to $\frac{3}{4}$ of E.'s, which is 3 times $\frac{2}{3}$ of D.'s, being on interest for 8 years, at 5 per cent., gives \$800 interest. What is the fortune of each?

42. The interest on A.'s, B.'s and C.'s fortunes, for 5 years, at 8 per cent., is \$1040. What is the fortune of each, provided they are to each other as $\frac{1}{2}$, $\frac{1}{3}$, and $\frac{1}{4}$?

43. The interest of A.'s, B.'s, and C.'s fortunes, for $5\frac{1}{3}$ years, at 6 per cent., is \$800. What is each of their fortunes, provided B.'s is twice A.'s, and B.'s and C.'s are equal?

44. A.'s fortune added to $\frac{2}{3}$ of B.'s, which is to A.'s as 2 is to 3, being put on interest for 6 years, at 4 per cent., amounts to \$124. What is the fortune of each?

45. D.'s money added to 4 times E.'s, which is equal to D.'s, being on interest for 10 years, at 5 per cent., amounted to \$3000. What was each of their fortunes?

46. The sum of $\frac{2}{3}$ of A.'s and $\frac{1}{2}$ of B.'s money, being on interest for 8 years, at 5 per cent., amounts to \$2100. Provided $\frac{1}{2}$ of B.'s money is twice $\frac{2}{3}$ of A.'s, how much money has each?

47. $\frac{2}{5}$ of the cost of C.'s house, increased by $\frac{3}{5}$ of the cost of his farm, being placed on interest for 10 years, at 7 per cent., amounts to \$17000. What is the cost of each, if $\frac{2}{5}$ of the cost of the house is only $\frac{1}{4}$ of $\frac{3}{5}$ of the cost of the farm?

48. If $\frac{5}{7}$ of A.'s fortune in 2 years and 4 months, at 6

per cent., amounts to $570, what is his whole fortune ?

49. The sum of A.'s and B.'s fortunes, in 4 years and 8 months, at 6 per cent., amounted to $256. What was each of their fortunes, provided $\frac{2}{3}$ of A.'s fortune equaled B.'s ?

50. The interest for 5 years, at 6 per cent., on $\frac{2}{3}$ of the money Morgan owes is $180; and the interest for the same time and rate per cent., on $\frac{2}{5}$ of the money due him, is $120. How much has Morgan after paying his debts ?

51. The money John paid for a sheep, a cow, and a horse, in 8 years, at 10 per cent., would give such an interest, as would in $\frac{3}{4}$ as long, at $\frac{1}{2}$ as great a per cent., amount to $104; how much did he pay for each, provided the sheep cost $\frac{1}{2}$ as much as the cow, and the cow $\frac{1}{3}$ as much as the horse ?

52. The interest of the sum of $\frac{1}{2}$ of Simpson's, $\frac{5}{9}$ of Eyer's, and $\frac{5}{12}$ of Domer's fortunes, for 3 years 7 months and 6 days, at 10 per cent., is such as will in the same time, at $\frac{1}{2}$ the rate per cent., amount to $531. What is the fortune of each, provided $1\frac{1}{2}$ times Domer's part of the principal equals $\frac{3}{4}$ of Eyer's, and $\frac{7}{10}$ of Eyer's part of the principal equals $\frac{1}{5}$ of Simpson's ?

53. The interest of the sum of $\frac{1}{2}$ of A.'s, and $\frac{2}{3}$ of B.'s fortune, for a *certain* time, at 2 per cent., was to *this sum* as 9 to 250. And the amount of *this interest* for 25 times as long, at 10 times as great a per cent., was $180. What was each of their fortunes, provided A.'s fortune was to B.'s as 1 to 3 ? And how long was the first on interest ?

REMARK.—Since the interest was to the principal as 9 to 250, $\frac{9}{250}$ of the principal equals the interest. Hence, 1 year 9 months and 18 days is the time required, &c.   . .   `

NUMERATION AND DECIMALS.

In view of the importance and utility of the Decimal system of Weights, Measures, and Currency, the Notation of our Decimal Arithmetic, including Decimal Fractions, is now presented, that pupils may comprehend the simplicity of the system, and more easily perform calculations on the Tables.

The Decimal System of Measures and Weights having the Metre, the Litre, and the Gram as principal units, and of currency having the Dollar as the principal unit in the United States, Canada, and some other countries, and the Franc as the principal unit in France, will be found simple and useful to insure accuracy of computation and economy of time.

The denominations are also systematic. For those greater than the principal unit, prefixes from the Greek language are used, as follows: *deka* (from *deka*, ten); *hecto* (from *ekaton*, hundred); *kilo* (from *kilion*, thousand); and *myria* (from *murias*, ten thousand). For denominations less than the principal unit, prefixes from the Latin language are used, as follows: *deci* (from *decem*, ten); *centi* (from *centum*, hundred); *milli* (from *mille*, thousand), etc. Therefore, 1 decimetre is equal to one tenth part of a metre, and 1 decametre is ten metres. See Tables on pages 55, 56.

Figures express numbers and parts of numbers.

A *Whole Number* is called an Integral Number, and the figures expressing it are named *Integers;* as, 1, 8, 25.

A *Part of a Number* is called a Fractional Number; and the figures expressing it are named *Fractions;* as, $\frac{1}{2}$, $\frac{11}{36}$, .2, or $\frac{2}{10}$.

The whole number of which the fraction is a part, is called the unit of the fraction.

A *Common Fraction* is one or more of the equal parts of any quantity, and is expressed by two terms, the Numerator and Denominator, with a line between them, as $\frac{1}{2}$, $\frac{2}{3}$ of a yard, $\frac{4}{5}$ of 10.

A *Decimal Fraction* is a division of a unit into tenths, etc., according to the scale of the Numeration

8

Table, and is expressed by its numerator only, having a dot (.), called a decimal point, before it; thus, $.1 = \frac{1}{10}$, $.23 = \frac{23}{100}$.

A *Mixed Number* is a number expressed by an *Integer* and a *Fraction*; as, $3\frac{1}{2}$, $20\frac{2}{7}$.

A *Mixed Fraction* is a decimal and common fraction united in one; thus, $.3\frac{1}{2}$ is read three and one half tenths.

An *Abstract Number* is a number that expresses no particular kind of quantity; or number considered without denominative quantity; as 2, 19, $7\frac{1}{2}$.

A *Denominate Number* is a number used in connection with a name of quantity; as, 4 yards, $8\frac{1}{3}$ bushels, $24, 40 metres.

In the *Decimal System of Notation,* ten characters, or figures, are used, viz., the nine significant figures, 1, 2, 3, 4, 5, 6, 7, 8, 9, and the cipher, 0.

Each figure may have two values, one when it is alone, which is its simple value; the other when it is used with other figures, which is its local value: thus, 1 represents one, or a unit; 4 represents four, or four units.

If these two figures are put together, one changes its value; thus, 14 represents *fourteen*, or one ten and four units; and 41 represents *forty-one*, or four tens and one unit.

If another figure be added on the right, both of the two figures change in value ten-fold; thus, 142 signifies one hundred and forty-two; and 410 is four hundreds one ten and no units, or four hundred and ten.

If another figure be added on the left, it represents its proper local value, but does not affect the figures to which it is thus added, because it has not changed their places; thus, 3142 represents three thousands, one hundred, four tens, and two units, or is read, three thousand one hundred and forty-two.

The place of a figure denotes its value, according to the scale of the

NUMERATION TABLE.

Billions.	Hundred millions.	Ten millions.	Millions.	Hundred thousands.	Ten thousands.	Thousands.	Hundreds.	Tens.	Units.	
									2	Two.
								2	3	Twenty-three.
							2	3	5	Two hundred thirty-five.
						2	3	5	4	Two thousand three hundred fifty-four.
					2	3	5	4	6	Twenty-three thousand five hundred forty-six.
				2	3	5	4	6	0	Two hundred thirty-five thousand four hundred sixty.
			2	3	5	4	6	0	8	Two millions three hundred fifty-four thousand six hundred and eight.
		2	3	5	4	6	0	8	1	Twenty-three millions five hundred forty-six thousand and eighty-one.
	2	3	5	4	6	0	8	1	9	Two hundred thirty-five million four hundred sixty thousand eight hundred and nineteen.
2	3	5	4	6	0	8	1	9	7	Two billions three hundred fifty-four millions six hundred and eight thousand one hundred ninety-seven.

NOTE.—These denominations are sometimes divided into periods of three each from the right, the first period being of units, the second being of thousands, the third being of millions, the fourth being of billions, the fifth being of trillions, the sixth being of quadrillions, etc.

A figure being moved one place to the left, in the above table, increases its value ten-fold, and ten of one place equal one of the next place on its left.

Although a cipher expresses no number, it is used to locate figures of other orders in their proper places; thus, to express six thousand and eighty-one (6081), the cipher locates the figure 6 in its proper place.

Read the following numbers:

302 — 4516 — 2007 — 180960 — 81108 — 9716439 — 17647 — 2854 — 430000 — 5000006 — 340657 — 9977700 — 18001800 — 24697 — 684321 — 5402835.

The Notation of Decimal Fractions or Decimals, is in harmony with that of Integers, as shown in the following

TABLE OF INTEGERS AND DECIMALS.

Etc.	Millions.	Hundred thousands.	Ten thousands.	Thousands.	Hundreds.	Tens.	Units.	decimal point.	Tenths.	Hundredths.	Thousandths.	Ten thousandths.	Hundred thousandths.	Millionths.	Etc.
				INTEGERS.				.			DECIMALS.				
Two tenths,								.	2						
Twenty-three hundredths,								.	2	3					
Two hundred and thirty-five thousandths,								.	2	3	5				
Three hundred and fifty-six ten thousandths,								.	0	3	5	6			
Twenty-one thousand and four hundred thousandths,								.	2	1	0	0	4		
Seven hundred and forty millionths,								.	0	0	0	7	4	0	

The figures in a decimal expression are the numerator of a decimal fraction; its denominator is understood to be 1 followed by as many ciphers as there are decimal places in the expression; thus, the decimal fraction, .0356, when written as a common fraction, is $\frac{356}{10000}$.

The denominator of a decimal fraction is always equal to 1 unit, though it may be understood as tenths, hundredths, etc.

Whole numbers, or integers, and decimals, can be written together in one line of figures, by using the decimal point to separate them, because there is the same scale of increasing or decreasing value with both; thus, 2.2 is read, two and two tenths, and signifies 2 units and $\frac{2}{10}$ of another unit; 431.56 is read, four hundred and thirty-one and fifty-six hundredths of one.

The Addition, Subtraction, Multiplication, and Division of Mixed Decimals, or Numbers expressing integers and decimals, are the same as in whole numbers; thus,

ADDITION.	SUBTRACTION.	MULTIPLICATION.	DIVISION.
43.2	43.20	43.2	.24)43.20(180 *Quotient.*
.24	.24	.24	24
43.44 *Sum.*	42.96 *Rem.*	1728	1920
		864	1920
		10.368 *Product.*	*No Rem.*

NOTE.—In operations in Addition and Subtraction, the figures of each denomination of one number must be added to or subtracted from the figures of the same denominations in the other numbers; thereby, the decimal points will be under each other, and before the tenths placed in the *Sum* and *Remainder*.

In Multiplication, the *Product* must have as many decimal places as there are decimal places in both the Multiplicand and Multiplier. If necessary, ciphers must be prefixed to the product to make its number of decimal places equal.

In Division, the number of decimal places in the *Quotient* must equal the decimal places in the Dividend diminished by the number of decimal places in the Divisor. If necessary to produce this result, ciphers must be prefixed to the Quotient, or affixed to the Dividend, and when both Dividend and Divisor contain the same number of decimal places, the Quotient is a whole number, with or without a Remainder, as the case may be.

Annexing a Cipher on the right of a decimal expression does not alter its value, but multiplies both terms of the fraction by 10, and reduces it to the next lower denomination; thus, $.2 = \frac{2}{10} = .20 = \frac{20}{100}$, or $= .200 = \frac{200}{1000}$.

A decimal, or a mixed number of integers and decimals, can be multiplied by 10, 100, 1000, etc., by removing the decimal point as many places to the right as there are ciphers in the multiplier.

Multiply .678 by 10, by 100, by 1000; 178.9 by 10; 19.320 by 10; $16.0054 \times 100 = ?$ $54.37898 \times 10000 ?$ $732.98 \times 1000 = ?$

Prefixing a Cipher in a decimal expression, divides the decimal by 10, by removing each of its figures one place farther from the decimal point; thus, $.5 = \frac{5}{10}$, $.05 = \frac{5}{100}$, $.005 = \frac{5}{1000}$.

A decimal, or a mixed number of integers and decimals, can be divided by 10, 100, 1000, etc., by removing the decimal point as many places to the left as there are ciphers in the divisor.

Divide 13.41 by 10, by 100, by 1000; 476.9 by 100, by 10000; .17 by 10; .176 by 1000; $438.15 \div 10 = ?$ $167.009 \div 10 = ?$ $\div 1000 = ?$

.75 of a metre = how many decimetres and centimetres? 28.03 expresses how many metres, decimetres, and centimetres?

Divide 28.03 by such use of the decimal point, so that it will express 2 grammes 8 decigrammes and 3 milligrammes. Express 3650 cents in dollars and cents, by the use of the decimal point.

Read 45.75 in United States currency, in French currency, in decimal measures of weight, decimal measures of capacity. Read 196.071.

A Common Fraction can be changed to an equivalent Decimal Fraction, by dividing its numerator with ciphers annexed, by its denominator, and making as many decimal places in the quotient as there are ciphers annexed; thus, $\frac{1}{2} = 10 \div 2 = 5$.

Reduce $\frac{3}{8}$ to its equivalent decimal.

ANALYSIS 1ST.—3 = 30 tenths÷8 gives 3 tenths and 6 tenths remaining; 6 tenths = 60 hundredths ÷ 8 gives 7 hundredths and 4 hundredths remaining; 4 hundredths = 40 thousandths ÷8 gives 5 thousandths and no remainder; hence $\frac{3}{8}$ = .375.

ANALYSIS 2ND.—$\frac{3}{8} = \frac{3.000}{8.000} = \frac{.375}{1.000} = \frac{.375}{1} = .375$.

$\frac{1}{2}$ of $1 equals how many cents?

.5 of $1 equals how many cents?

What is the decimal expression of $\frac{4}{5}$?

.8 of $1 equals how many dimes?

$\frac{3}{4}$ or .75 of $1 equals how much money?

$\frac{3}{4}$ of a metre equals how many decimetres and centimetres?

QUESTIONS, QUERIES, AND PUZZLES

FOR

PUPILS AT HOME.

————◆·◆————

1. A hound is in pursuit of a fox that is 10 rods ahead of him, and *while* the hound runs 10 rods the fox runs 1 rod, (*i. e.* while the hound runs a certain distance, the fox runs one-tenth of that distance.) Will the hound overtake the fox? The conditions remaining the same, what is the greatest distance they can run?

2. A hound is in pursuit of a fox that is 10 rods ahead of him, and *while* the fox runs 1 rod the hound runs 10 rods. How far will the hound run before he overtakes the fox?

3. Place four 5's in such a position that they shall equal 6½.

4. A boy was sent to a spring with a 5 and a 3 quart measure to procure exactly 4 quarts of water. How did he measure it?

5. What is the difference between twice 25, and twice 5 and 20?

6. A man had 9 pigs and put them in 4 pens, with an odd number of pigs in each pen. How did he divide them?

7. Two men have an 8 gallon cask full of wine, which they desire to divide equally between them. How can they effect this division, provided the only measures they have are a 5 gallon cask and a 3 gallon cask?

8. Place four 2's in such a manner that they shall exactly equal 23.

9. Place 9 apples in 10 rows so that each row shall contain 3 apples.

10. A squirrel finding 9 ears of corn in a box, took from it daily, 3 *ears;* how many days was he in removing the corn from the box?

11. If from *six* you take IX, and from IX you take ten; and if fifty from forty be taken, there will then just half a dozen remain.

12. Edward, Maria, and their mother went to market. Edward had 60 apples, and sold them, at 2 for 1 cent; Maria had 60 apples, and sold them, at 3 for 1 cent. Their mother had 120 apples, and sold them, at the rate of 5 for 2 cents. Which received the most, the children or the mother, and why?

13. A gentleman desiring to see an inmate of a prison, was asked by the keeper whether he was related to the culprit, replied: "Brothers and sisters have I none, but his father is my father's son." What relation was the gentleman to the prisoner?

14. A man having a fox, a goose, and a peck of corn, was desirous of crossing a river. He could take only *one* across at a time, and if he left the fox and goose, while he took the corn over, the fox would kill the goose; but if he left the goose and corn, the goose would eat the corn. How shall he get them all safely across the river?

15. A snail wants to get up a wall 20 feet in height, during the day it climbs 5 feet, but slips back 4 feet every night; how many days would it take to reach the top?

16. A man purchased a hat for $5, and handed the merchant a $50 bill to pay for it: the merchant being unable to make the change, sent the bill to a broker, got it changed, and then gave the man who bought the hat $45. The broker, after the purchaser of the hat had gone, discovered that the bill was counterfeit, and, therefore, returned it to the merchant and received $50 good money. How much did the merchant lose by the operation?

17. Place ten pennies in a row, then carry one over two, leaving it upon the third, and continue doing this until the ten pennies occupy only five places, with two in each place.

18. Two boys laid a wager as to which could lift the most. One lifted ninety-nine pounds, and the other a hundred, and *won*. How many pounds did both lift?

19. A frog, at the bottom of a well 10 feet deep, ascends 3 feet every jump. How many jumps must he take to get out?

20. A drover being asked how many horses he had, replied, "My horses together have twenty *fore* legs;" how many horses had he?

21. Write 12 in such a manner, that you can show its half to be 7.

22. "I am constrained to plant a grove
      To please the lady that I love,
      This ample grove is to compose
      Nineteen trees in nine straight rows;
      Five trees in each row I must place,
      Or I shall never see her face."

www.ingramcontent.com/pod-product-compliance
Lightning Source LLC
Chambersburg PA
CBHW021805190326
41518CB00007B/456